LIE GROUPS: HISTORY, FRONTIERS AND APPLICATIONS

VOLUME VI

THE 1976 AMES RESEARCH CENTER (NASA) CONFERENCE ON

THE GEOMETRIC THEORY OF NON-LINEAR WAVES

EDITED BY

ROBERT HERMANN

ARTICLES BY: J. Corones, F. Estabrook, R. Gardner, R. Hermann, H. Morris, A. Scott, and H. Wahlquist.

MATH SCI PRESS
53 Jordan Road
Brookline, Mass. 02146 (USA)

Copyright © 1977 by Robert Hermann
All rights reserved

ISBN 0-915692-19-8

MATH SCI PRESS
53 Jordan Road
Brookline, Mass. 02146 (USA)

PRINTED IN THE UNITED STATES OF AMERICA

PREFACE

During the week of June 16-23 there was a small but intensive conference at Ames Research Center (NASA) concerning the application of differential geometry to the theory of nonlinear waves and solitons. Although there have been many conferences devoted to nonlinear waves in recent years, this was the first that emphasized the geometric side of the theory and that brought together most of the active workers in this subfield. Since this approach is now in full development--and promises many interesting and important applications--we decided to publish a short Proceedings, emphasizing semi-expository material which could serve as a guide to this approach for a wider audience.

An interesting special feature was that the mini-conference was embedded within a larger conference, "Differential and Algebraic Geometry for the Control Engineer" (whose Proceedings will also soon be published by Math Sci Press). Clyde Martin and I did this because we believe that the "geometric" control theory and nonlinear wave groups have much in common in terms of methodology and mathematics. Indeed, this aspect was very successful, and there was considerable communication and interaction between the two. Of course, this fits in well with my guiding philosophy in my own work for the "Interdisciplinary" series that there is a reasonably unified methodology for the application of "modern" differential geometry to a wide spectrum of problems in science and engineering.

Important features of the conference (in my view) were the heightened understanding we derived of the notion of "Bäcklund transformations" (in fact, Hugo Wahlquist's remark that they were special sorts of "prolongations" played a key role in my subsequent new work) and Hedley Morris' beginning calculations towards adapting the differential form-prolongation ideas to nonlinear wave equations in more variables and toward investigating the influence of Bäcklund transformations on topological properties. The centerpiece was the explanation by Frank Estabrook and Hugo Wahlquist of their "prolongation" method for studying nonlinear wave equations. Since their original paper, it has been recognized that their "prolongations" are what differential geometers call "connections", and thus there are especially close relations between their theory and modern differential geometry. In fact, this relation exhibits a contact (which so far is only via the similarity in the mathematics) with the theory of what physicists call gauge or Yang-Mills fields, which also involve the theory of connections. However, I believe the chief scientific significance of their work is that it is roughly as powerful as the better known Inverse Scattering Technique for the standard examples (Korteweg-de Vries, Sine-Gordon, etc.) but provides systematic methods for treating more complicated nonlinear wave equations (even those in more independent variables) where the latter technique becomes very chancey. In terms of that great cliche (which is profoundly true here), there is need for further research!

The lectures by J. Corones, R. Gardner, and A. Scott presented here also serve the function of introducing the reader to certain aspects of this broad "geometric" point of view. Corones provides a valuable link between the Estabrook-Wahlquist approach and the traditional methods. I share his belief that looking into the more traditional way for "integrability conditions" is often a remarkably simple, direct, and definitive approach to the problem. Gardner provides us with a survey of his work on shock waves using differential geometry, and "explicates" some very relevant and important (but little known) work by Cartan on nonlinear partial differential equations. (He has since extended these methods to study "Bäcklund transformations", a development which is of great potential importance.) Finally, Al Scott did much in his conversation to tie our work in with the rest of the "solitoners"--his short description of the alternate approach to Inverse Scattering is very interesting to me as a more "geometric" version that might tie in with our work. Kent Harrison described (but did not write up) his efforts to discover a "prolongation" structure for some of the nonlinear systems that appear in General Relativity. We all admired his courage and fortitude! Bob Kiehn told us informally about some fascinating possibilities for applying Solitons to fluid physics. Mike Hazewinkel gave us beautiful lectures on recent Soviet work interrelating the Korteweg-de Vries equations and algebraic geometry. Overall, this was the most enthusiastic and inspiring group of people I have ever been with at a conference, and I would like to wholeheartedly thank them.

In the forefront of our minds while planning the conference was the seemingly unfashionable idea that mathematical concepts and tools still have an important role to play in science and technology. We found it very useful and inspiring to hold such a conference in a "practical" laboratory like Ames Research Center. We are very grateful to NASA for providing the funds for the conference, and hope that their faith in our efforts can be repaid by theoreticians making contributions to the solutions of important practical problems In particular, it was in the back of our minds while planning this mini conference that our efforts towards understanding "nonlinear waves" (and Solitons, in particular) may be useful

PREFACE

in attacking the nonlinear problems that are so prominently present in aeronautics.

We are indebted to Clyde Martin, the organizer of the whole conference, and to our Host, Brian Doolin of the Ames Research Center who has wholeheartedly and generously supported our work. I would also like to thank F.K. Schwenk of NASA Headquarters, who provided us with the supplement to our basic funding that enabled us to hold this conference-within-a-conference, Joyce Martin, the conference secretary, and Karin Young, who typed so well.

The conference was supported by NASA Grant No. NSG-2148.

TABLE OF CONTENTS

	Page
PREFACE	iii

DIFFERENTIAL-GEOMETRIC PROLONGATIONS AND BÄCKLUND TRANSFORMATIONS. *Frank Estabrook, Hugo Wahlquist and Robert Hermann* 1

1. Introduction 1
2. The Basic Algebraic and Differential Operations 5
3. The Natural Operations in Differential Forms 7
4. Behavior of Differential Forms under Mappings and Changes of Variables. Integration. Stoke's Formula. Conserved Currents 15
5. Exterior Differential Systems 21
6. Conservation Laws of Differential Equations and Exterior Differential Systems 30
7. Conservation Laws and Prolongations 33
8. Multiple Pseudopotentials and Cartan-Ehresmann Connections 37
9. Multiple Prolongations Defined by Cartan-Ehresmann Connections with a Structure Group 41
10. Linear Pseudopotentials and the Inverse Scattering Equations 43
11. The "Inverse Scattering" Formalism as a "Complete Integrability" Condition, Leading to the Lax Equations. A Speculation about the Inverse Scattering Equation in Higher Dimensions 45
12. Specialization to the Korteweg-de Vries-Schrödinger Operator 48
13. The Bäcklund Transformations Described in Terms of Prolongations 54

Bibliography 62

Recent Literature of New Methods of Differential Geometry Applied to Nonlinear Partial Differential Equations and Soliton Theory 64

TABLE OF CONTENTS

Page

USING PSEUDOPOTENTIALS. *James Corones* — 67

    1. Introduction — 67
    2. A Brief Review of Pseudopotentials — 69
    3. An Example — 77
    4. Space-Time Dependent Pseudopotentials — 83

    Bibliography — 87

DIFFERENTIAL GEOMETRIC VIEWPOINTS ON THE DEVELOPMENT OF SHOCK WAVES. *Robert B. Gardner* — 89

    Bibliography — 104

BÄCKLUND TRANSFORMATIONS AND THE SINE-GORDON EQUATION. *Hedley C. Morris* — 105

    1. Prolongations — 105
    2. Bäcklund Transformations — 109
    3. Topological Charge [11] — 113
    4. The Massive Thirring Model — 118

    Bibliography — 123

PROLONGATION STRUCTURES OF NONLINEAR EVOLUTION EQUATIONS IN TWO AND THREE SPATIAL DIMENSIONS. *Hedley C. Morris* — 125

    1. Generalizing a Known Case — 126
    2. An Extended Ideal — 129
    3. The Kadomtsev-Petviashvili-Dryuma Equation — 131
    4. A Generalized Nonlinear Schrödinger Equation — 135
    5. Three Dimensions — 143

    Bibliography — 146

A PLAUSIBILITY ARGUMENT FOR THE MARCHENKO EQUATION *Alwyn C. Scott* — 147

    Bibliography — 153

DIFFERENTIAL-GEOMETRIC PROLONGATIONS
AND BÄCKLUND TRANSFORMATIONS

Frank Estabrook and Hugo Wahlquist
Jet Propulsion Laboratory
Pasadena, California
and
Robert Hermann
Harvard University
Cambridge, Massachusetts

1. INTRODUCTION

Research in the 19-th century on differential equations was quite different from that of today, since it was oriented towards finding specific sorts of solutions or doing certain calculations motivated by practical, geometric, or scientific considerations. The general directions of research were much more what we would call geometric, or group-theoretic, than today's work. Of particular interest to us is that many of the mathematical ideas that are now proving so useful in the study of nonlinear waves and solitons were developed in that period, especially the notion of Bäcklund transformation, which probably is the characteristic mathematical feature associated with solitons as physical phenomena.

Much of this 19-th century work has a characteristic flavor that has been completely lost. (This is remarkably different from other areas of mathematics or science, where by and large the 19-th century ideas have been assimilated into contemporary thought.) One has only to pick up books or articles by Bianchi, Darboux, Goursat, Lie, Riquier, Vessiot, Von Weber,... to appreciate the extent to which this material has become incomprehensible and lost.

Although differential geometry as a separate discipline did not really exist at that time, we can see now that this 19-th century work really involves what we would call a differential geometric (as opposed to an "analytic" or "algebraic") theory of differential equations. It is especially interesting that differential geometric methods are not so strongly limited to linear equations as those that are now used in science. The 19-th century mathematicians mentioned above were able to discover profound and fascinating facts about nonlinear equations which have no counterpart in current mathematical research.

It is very evident that the current scientific scene--both "pure" and "applied"--badly needs new mathematical methods for dealing with nonlinear differential equations.

For example, much of a subject like quantum field theory seems to be an attempt to extend the traditional "analytic" and "linear" methods of mathematical physics to differential equations which are intrinsically nonlinear, with little thought applied to the question of possible new geometric methods. Recent progress in experimental particle physics poses even more pressing questions of this sort, since it seems to suggest that the so-called gauge fields are fundamental objects; and they are especially strongly linked to geometric ideas. In addition, the study of such subjects as gravitation and turbulence has largely been outside of the mainstream of work in mathematical physics because of their strongly nonlinear nature.

The area of contemporary science where nonlinear problems have been tackled in a direct and forceful way is the theory of waves, mainly in one spatial dimension. (The recent book by Whitham [1] is an excellent source of information in this area.) This has been approached via the traditional methods of mathematical physics and applied mathematics: asymptotic and perturbation expansions, variational principles, numerical simulation, reduction to ordinary differential equations, etc. That these methods have given so much insight is a considerable tribute to their continuing vitality! The theory of "solitons" or "solitary waves" arises partly from this area, and partly (via the

Sine-Gordon equation) from relativistic field theory; it has recently begun to be intensely interesting and important, influencing research over a wide spectrum of areas in physics, engineering, and mathematics.

It has been evident that this work in nonlinear waves and solitons is directly linked to the 19-th century work on differential equations. Thus, further progress on those scientific and technological problems involving nonlinear differential equations might be linked to a reworking of the 19-th century material. (Think of the tidbits that have been developed so far as the gold dust from the stream--the Mother Lode is still to be found!)

We believe that the best way to understand this 19-th century work is through Elie Cartan's methods, which, in the modern mathematical literature, are now developed in coordinate free terms involving differentiable manifolds, differential forms, vector fields, etc. We are interested more in specific calculations than in proving general results (as in the mathematical literature), and we have, in fact, found that Cartan's methods are--once we became used to them--very useful for these calculations. We also believe that there is great potential for doing these calculations on a computer, for example, using the MACSYMA system; also in using them in problems where the system must be discretized.

We will obtain another fringe benefit from using Cartan's methods--a substantial new insight into the geometric nature of the equations (Korteweg de Vries, Sine-Gordon, nonlinear Schrödinger,...) which do admit soliton solutions; namely, we will show that their properties are closely related to those of certain connections in fiber bundles, a concept that Cartan himself was the first to isolate and study in adequate generality. In fact, Gauge Fields themselves are "connections", which indicates to us that there are probably strong links, at least at a mathematical level, between them and the nonlinear wave equations which admit solitons.

2. THE BASIC ALGEBRAIC AND DIFFERENTIAL OPERATIONS

Let $x = (x^1,\ldots,x^n) = (x^i)$, $1 \leq i,j \leq n$ be coordinates of a space. (In order to explain the basic ideas, we shall mainly work within the context of classical tensor analysis and mathematical physics.) A zero-form is a function $x \to f(x)$, i.e., a "scalar field". A one-form is an expression of the form:

$$\theta = a_i dx^i . \qquad (2.1)$$

(a_i) is a one-covariant tensor, i.e., a one-form is identical with a one-covariant tensor. (We use the summation convention in the usual way.) Writing it in the form (2.1) enables us to keep track of the transformation law of covariant vectors automatically. For if

$$(y^i)$$

is a new coordinate system,

$$dx^i = \frac{\partial x^i}{\partial y^j} dy^i ,$$

$$\theta = \left(a_i \frac{\partial x^i}{\partial y^j}\right) dy^j ,$$

i.e.,

$$a \to a \frac{\partial x}{\partial y}$$

is the transformation law.

A two-form is, from the tensor analysis point of view, an autosymmetric <u>two-covariant tensor</u> (a_{ij}). It is convenient to write it as

$$\omega = a_{ij} dx^i \wedge dx^j ,$$

because the transformation law on change of coordinates follows automatically. Of course, we must understand what is meant by "$\wedge$". It stands for an algebraic operation we describe below, called <u>exterior multiplication</u>, first described by Grassman in the 19-th century. Similarly, a three-form is of the form

$$\Omega = a_{ijk} dx^i \wedge dx^j \wedge dx^k ,$$

with completely skew-symmetric components (a_{ijk}). In general, an m-form is a **skew-symmetric, m-covariant tensor field**.

These differential forms admit several differential and algebraic operations which are "natural" in the sense that they are invariant under **arbitrary** change of coordinates. (Thus, if $\theta = a_i dx^i$, the operation

$$\theta \rightarrow \frac{\partial \theta}{\partial x^1} = \left(\frac{\partial a_i}{\partial x^1}\right) dx^i ,$$

differentiation of the components by the first coordinate, is **not** natural.) The prime feature in Cartan's work is that **differential equations are to be written in a form which only involves these natural operations**. This feature implies a completely different look and feel to the traditional material of engineering and physics, which, of course, has been a barrier to the effective utilization of Cartan's methods by most engineers and physicists.

3. THE NATURAL OPERATIONS IN DIFFERENTIAL FORMS

First, we deal with those of "algebraic" type, i.e., which do not involve derivatives.

A) $(\theta_1, \theta_2) \to \theta_1 \wedge \theta_2$: <u>exterior multiplication</u>

a) $\theta_1 = f$, $\theta_2 = g$, zero-forms, then $\theta_1 \wedge \theta_2 = fg$, the usual product of functions.

b) $\theta_1 = f$, a zero-form, $\theta_2 = a_i dx^i$, a one-form $\theta_1 \wedge \theta_2 = (fa_i)dx^i$, i.e., multiply the components. Thus, in tensor analysis terms, this involves simple multiplication by a scalar field.

c) $\theta_1 = a_i dx^i$; $\theta_2 = b_i dx^i$

$\theta_1 \wedge \theta_2 = \frac{1}{2}(a_i b_j - a_j b_i)dx^i \wedge dx^j$

d) $\theta_1 = a_i dx^i$; $\theta_2 = b_{ij} dx^i \wedge dx^j$

$\theta_1 \wedge \theta_2 = \frac{1}{3}(a_i b_{jk} + a_k b_{ij} + a_j b_{ki} dx^i \wedge dx^j \wedge dx^k$

Rules:

$$\theta_1 \wedge (\theta_2 + \theta_3) = \theta_1 \wedge \theta_2 + \theta_1 \wedge \theta_3$$

$$\theta_1 \wedge \theta_2 = (-1)^{pq} \theta_2 \wedge \theta_1$$

if degree $\theta_1 = p$, degree $\theta_2 = q$,

$$\theta_1 \wedge (\theta_2 \wedge \theta_3) = (\theta_1 \wedge \theta_2) \wedge \theta_3$$

PROLONGATIONS   9

B)   Contraction or inner product ⌟.

Let $A = (A^i)$ be a one-contravariant tensor. For geometric reasons explained below ("Lie derivative"), we denote A as a first order linear differential operator

$$A = A^i \frac{\partial}{\partial x^i} ,$$

and call it a <u>vector field</u>.

$$A \lrcorner \theta \equiv \underline{\text{contraction of}} \;\; \theta \;\; \underline{\text{by}} \;\; A$$

If $\theta = a_i dx^i$,

$$A \lrcorner \theta = A^i a_i \equiv \text{a scalar field} ;$$

$\theta = a_{ij} dx^i \wedge dx^j$,

$$A \lrcorner \theta = 2 A^i a_{ij} dx^j ,$$

a one form.

$$\theta = a_{ijk} dx^i \wedge dx^j \wedge dx^k ,$$

$$A \lrcorner \theta = 3 A^i a_{ijk} dx^j \wedge dx^k ,$$

and so forth.

Here is the main algebraic rule relating contraction and exterior multiplication:

$$\boxed{A \lrcorner (\theta_1 \wedge \theta_2) = (A \lrcorner \theta_1) \wedge \theta_2 + (-1)^p \theta_1 \wedge (A \lrcorner \theta_2)}$$

$$\text{where } p = \text{degree } \theta_1$$

Now we list the natural differential operators.

C) <u>Exterior derivative.</u>

$$f \text{ a zero-form: } df = \frac{\partial f}{\partial x^i} dx^i$$

$$\theta = a_i dx^i \rightarrow d\theta = da_i \wedge dx^i$$

$$= \frac{1}{2} \left(\frac{\partial a_i}{\partial x^j} - \frac{\partial a_j}{\partial x^i} \right) dx^i \wedge dx^j$$

$$\omega = a_{ij} dx^i \wedge dx^j \rightarrow d\omega = da_{ij} \wedge dx^i \wedge dx^j$$

and so forth.

Rules:

$$d(\theta_1 + \theta_2) = d\theta_1 + d\theta_2$$

$$d(\theta_1 \wedge \theta_2) = d\theta_1 \wedge \theta_2 + (-1)^p \theta_1 \wedge d\theta_2$$

$$\text{if } p = \text{degree } \theta_1$$

D) <u>Lie derivative</u>.

$$A = A^i \frac{\partial}{\partial x^i} \quad \text{a vector field}$$

θ a p-form

$\mathscr{L}_A(\theta)$, the <u>Lie derivative of</u> θ <u>by</u> A, is again a p-form:

$$\mathscr{L}_A(f) = A^i \frac{\partial f}{\partial x^i} \equiv A(f) \quad .$$

Note: This is the reason for identifying a "vector field", i.e., a one-contravariant vector field, with a first order linear differential operator. We shall see that this identification is "natural" from the point of view of change of coordinates.

$$\theta = a_i dx^i \quad ; \qquad A = A^i \frac{\partial}{\partial x^i} \quad ,$$

$$\mathscr{L}_A(\theta) = A(a_i) dx^i + a_i d(A(x^i))$$

$$= \frac{\partial a_i}{\partial x^j} A^j dx^i + a_i \frac{\partial}{\partial x^j}(A^i) dx^i$$

$$\theta = a_{ij} dx^i \wedge dx^j$$

$$\mathscr{L}_A(\theta) = A(a_{ij}) dx^i \wedge dx^j + a_{ij} d(A^i) \wedge dx^j + a_{ij} dx^i \wedge d(A^j)$$

and so forth.

The Lie derivative is often (in the tensor analysis literature) called the "dragging along operation". Think of the vector field A as the velocity field of a flow

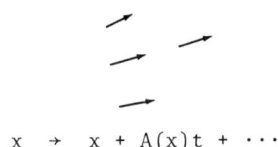

$$x \to x + A(x)t + \cdots$$

A(f) is then the derivative

$$\frac{\partial}{\partial t} f(x + A(x)t) \quad .$$

Similarly, $\mathscr{L}_A(\theta)$ is the first order change in θ when it is moved around by the transformation. In fact, for this reason Lie derivative is not only a "natural" operation on differential forms, i.e., skew-symmetric covariant derivatives, but on all types of tensor fields. In particular, its action

$$B \to [A,B]$$

on vector fields is given a special name--<u>Jacobi bracket</u>-- and plays an especially important role in differential geometry and Lie group theory.

Here is its formula in local coordinates: If

$$A = A^i \frac{\partial}{\partial x^i} \quad ; \qquad B = B^j \frac{\partial}{\partial x^j} \quad ,$$

$$[A,B] = A(B^j)\frac{\partial}{\partial x^j} - B^j\left[\frac{\partial}{\partial x^j}, A\right]$$

$$= A^i \frac{\partial B^j}{\partial x^i}\frac{\partial}{\partial x^j} - B^j \frac{\partial A^i}{\partial x^j}\frac{\partial}{\partial x^i}$$

$$= \left(A^j \frac{\partial B^i}{\partial x^j} - B^j \frac{\partial A^i}{\partial x^j}\right)\frac{\partial}{\partial x^i}$$

Of course, all these operations, when suitably specialized to three and four dimensions, lead to the natural vector analysis notation that physicists have long used--div, grad, cul, and all that. For further details, see the books on differential geometry listed in the Bibliography.

Some useful identities

$$\mathscr{L}_A(\theta_1 + \theta_2) = \mathscr{L}_A(\theta_1) + \mathscr{L}_A(\theta_2)$$

$$\mathscr{L}_A(\theta_1 \wedge \theta_2) = \mathscr{L}_A(\theta_1) \wedge \theta_2 + \theta_1 \wedge \mathscr{L}_A(\theta_2)$$

$$\mathscr{L}_A(d\theta) = d\mathscr{L}_A(\theta)$$

$$\mathscr{L}_A(B \lrcorner \theta) = [A,B] \lrcorner \theta + B \lrcorner \mathscr{L}_A(\theta)$$

$$\mathscr{L}_A(\mathscr{L}_B(\theta)) = \mathscr{L}_{[A,B]}(\theta) + \mathscr{L}_B(\mathscr{L}_A(\theta))$$

$$\mathscr{L}_A(\theta) = A \lrcorner d\theta + d(A \lrcorner \theta)$$

This last identity links three natural operations--and expresses Lie derivative in terms of exterior derivative and contraction--and is a key identity in many applications, particularly to the calculus of variation.

We have emphasized that differential forms are one sort of tensor field, which implies that they transform naturally under change of variables. However, they were described mathematically well before the notion of "tensors" was well understood. This reflects another feature--<u>their appearance in integral calculus as the expressions underneath integral signs</u>. Steve Weinberg, in his well known book on graviation [2], expresses distinct annoyance that mathematicians are hiding from him the fact that differential forms are "just" special types of tensor fields. Presumably, he (and other physicists) do not understand that this role as "integrands" has led to their use in topology and differential geometry (particularly in Cartan's work) in ways that completely transcend their being "merely" tensor fields! In fact, their characteristic role in integral calculus suggests that they should play a basic role when <u>physical laws are expressed in terms of integral rather than</u> (as usual) <u>in terms of differential relations</u>. For example, electromagnetism takes a remarkably simpler mathematical shape when expressed in terms of differential forms than in its traditional format in terms of three-dimensional vector analysis.

4. BEHAVIOR OF DIFFERENTIAL FORMS UNDER MAPPINGS AND CHANGES OF VARIABLES. INTEGRATION. STOKES' FORMULA. CONSERVED CURRENTS.

There are two ways of looking at what is meant by "invariance" in tensor analysis--the <u>passive</u> and <u>active</u>. The former is the usual one--"points" remain the same but the ways of labelling them change. In the latter, the "points" actually change by a "mapping". Usually, one supposes the "mappings" are <u>invertible</u>. However, differential forms have an "invariance" under mappings which are not necessarily invertible, and for which the domain and range have different dimensions.

Let X and Y be two spaces, (x^i) coordinates for X, (y^i) coordinates for Y. Let

$$\phi: Y \to X$$

be a mapping expressed in these coordinates by

$$\phi(y) = x(y) .$$

Differential forms map <u>backwards</u> (or <u>dually</u> to ϕ), i.e., to a differential form θ on X we assign a differential form, denoted as $\phi^*(\theta)$, on Y (but not necessarily conversely).

$$\theta = f, \text{ a zero-form on } X$$

$\phi^*(f)$ is $y \to f(x(y))$, a zero-form on Y

$$\theta = a_i dx^i$$

$$\phi^*(\theta) = \phi^*(a_i) d\phi^*(x^i) = a_i(x(y)) d(x^i(y))$$

$$= a_i(x(y)) \frac{\partial}{\partial y^j} (x^i(y)) dy^j$$

$$\phi^*(a_{ij} dx^i \wedge dx^j) = \phi^*(a_{ij}) d(x^i(y)) \wedge d(x^j(y))$$

etc.

Here are the general rules:

$$\phi^*(\theta_1 \wedge \theta_2) = \phi^*(\theta_1) \wedge \phi^*(\theta_2)$$

$$\phi^*(d\theta) = d\phi^*(\theta)$$

$$\phi^*(\theta_1 + \theta_2) = \phi^*(\theta_1) + \phi^*(\theta_2)$$

This invariance property--together with the skew-symmetry--leads to the <u>integral calculus</u> properties of differential forms mentioned in the previous section. First, suppose

$$\dim X = n \quad,$$

and θ is an n-form. The skew-symmetry forces θ to have only one degree of freedom, i.e., to be written as

$$\theta = f dx^1 \wedge \cdots \wedge dx^n \quad.$$

Thus,

$$\int_X \theta = \int f(x) \, dx^1 \cdots dx^n \quad .$$

(One verifies that this definition is <u>independent of change of coordinates</u>--this is basically a consequence of the change-of-variable-via-Jacobian rule of multiple integrals, and the fact that the Jacobian is a determinant, which, in turn, arises naturally via exterior algebra. Thus,

$$J = \det\left(\frac{\partial x^i}{\partial y^j}\right)$$

$$dx^1 \wedge \cdots \wedge dx^n = J dy^1 \wedge \cdots \wedge dy^n \quad .)$$

Now, suppose that θ is an m-form on X, dimension X = n > m. Let Y be an m-dimensional submanifold of X. Regard this as a mapping

$$\phi: Y \to X$$

which is one-one and such that the Jacobian matrix $(\partial x^i/\partial y^j)$ has rank m. Set:

$$\int_Y \theta \equiv \int_Y \phi^*(\theta) \quad .$$

This is called the <u>integral</u> of the m-form θ over the submanifold $\phi(Y)$.

Examples:

    $m = 1$ <u>Line integral</u>: Parameterize Y by t.

$$\phi(t) = x(t)$$

is a parameterized curve.

$$\theta = a_i dx^i$$

$$\int_Y \theta = \int a_i(x(t)) \frac{dx^i}{dt} dt$$

$$\equiv \text{Line integral in the classical sense}$$

    $m = 2$ <u>Surface integral</u>: Parameterize Y by (s,t)

$$\phi(s,t) = x(s,t)$$

$$\theta = a_{ij} dx^i \wedge dx^j$$

$$\int_Y \theta = \iint a_{ij}(x(s,t)) \begin{vmatrix} \frac{\partial x^i}{\partial t} & \frac{\partial x^j}{\partial s} \\ \frac{\partial x^i}{\partial s} & \frac{\partial x^j}{\partial t} \end{vmatrix} ds\, dt$$

<u>Stokes' Formula</u>:

$$\int_{\partial Y} \theta = \int_Y d\theta$$

where degree $\theta = (m-1)$, Y is an m-dimensional submanifold, and ∂Y denotes its boundary. (There are certain "orientation"

PROLONGATIONS

details which should be included, and that are more-or-less familiar to anyone who has assimilated advanced calculus.)

Stokes' formula is familiar to physicists as the general form of the integral formulas (Gauss, Green, Stokes, etc.) that play a role in three-dimensional vector analysis and, through that, in field-theoretic physics. It also plays the key role in the "topological" interpretation of differential forms ("de Rham's theorems"), and in the calculus of variations. For the purpose of this paper, its role is defining the concept of "conservation laws" is also critical.

Suppose we are dealing with a physical field theory with one space variable "x" and one time variable t. Suppose X has variables x,t and other variables y. Consider a two-dimensional region R in X which, in the (x-t)-plane, has the form of a "strip"

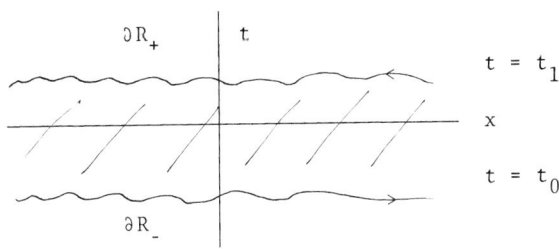

Let θ be a one-form in X, i.e., one involving the variables x, t, and y. Then, by Stokes' formula:

$$\int_R d\theta = \int_{\partial R} \theta = \int_{\partial R_+} \theta + \int_{\partial R_-} \theta$$

The region in X will be defined by giving y as a function $y(x,t)$ of these variables. Thus,

$$\int_R d\theta \quad \text{will equal zero if} \quad d(\theta(y(x,t))) = 0$$

Suppose, for example, that:

$$\theta = a(y)dx + b(y)dt \quad .$$

This means that:

$$\frac{\partial}{\partial t}(a(y(x,t))) = \frac{\partial}{\partial x}(b(y(x,t))) \quad .$$

Now, if $\int_R d\theta = 0$, then the integral of θ over the upper boundary equals (taking into account orientation) integral over lower line. Thus

$$\int a(y(x,t_1)) \, dx = \int a(y(x,t_0)) \, dx$$

This is a typical "conservation law", i.e., it says that the "space integral" is <u>independent of the time</u>. What we have done is to convert it into a <u>differential</u> condition: The pull back of $d\theta$ by the submanifold map is zero. We shall now see how this directly ties up with the point of view of Cartan's theory of exterior differential systems.

5. EXTERIOR DIFFERENTIAL SYSTEMS

As we have emphasized, differential forms are the geometric objects which have the properties--invariance under arbitrary mappings and integrands for multidimensional integrals--which make them suitable for description of physical and geometric phenomena. Since physical laws (and many geometric ones too!) are described by differential equations, it seems desirable to have a way to describe them in terms of differential forms. This was done by E. Cartan, leading to a theory which he called that of <u>exterior differential systems</u> [4,5,6].

Let X be a manifold. (It is convenient now to use the ideas and terminology of differential manifold theory. See [3,6,7]. The reader who is more accustomed to the geometric framework of the traditional engineering and physics literature can think of X as Euclidean space, with one or more prescribed coordinate systems. Our reason for thinking in terms of "manifolds" is not (as in topology) to make "global" formulations of the ideas--although that is a useful byproduct--but to achieve conceptual clarity. An <u>exterior differential system</u> on X --denoted, say, be ED -- is a collection of differential forms on X such that:

a) If $\theta_1, \theta_2 \in$ ED, then $\theta_1 + \theta_2 \in$ ED

b) If $\theta_1 \in$ ED, θ_2 is an arbitrary form on X, then $\theta_1 \wedge \theta_2 \in$ ED

c) $d\theta \in ED$ for each $\theta \in ED$.

A subset of ED is said to <u>generate</u> ED if the smallest exterior differential system containing that subset is ED itself.

A submanifold Y of X is said to be an <u>integral submanifold</u> of ED if each form θ in ED is zero when "pulled back" to the submanifold. (We also call the process of pulling a form back to a submanifold the <u>operation of sectioning the differential form</u>. Yet another usage is to say the differential form is <u>restricted</u> to the submanifold.) Precisely, if $\phi: Y \to X$ is the map which "parameterizes" the submanifold, then

$$\phi^*(\theta) = 0 , \qquad \text{for all } \theta \in ED . \qquad (5.1)$$

For example, suppose

$$\theta = a_i dx^i$$

is a one-form. Let y be the coordinate of Y, $y \to x(y)$ the functions that define the submanifold in parameterized form. (Of course, the submanifold may be described initially by relations of the form

$$f(x) = 0 .$$

It must be possible, in principle at least, to solve them as $y \to x(y)$, so that

$$f(x,y(x)) = 0 .$$

PROLONGATIONS

Of course, the Implicit Function Theorem gives conditions that assure one can do this in almost all conceivable practical situations.) The condition that (5.1) be satisfied is then that:

$$a_i(x(y)) \frac{dx^i}{dy} = 0 \quad .$$

These are differential equations for $x(y)$, so that the notion of "integral submanifold of an exterior differential system" is nothing but a notion of a "differential equation" in a fancier guise.

Of course, in this form the differential equations have what physicists call a "gauge invariance"--the change of coordinates y of the submanifold maps integral submanifolds into integral submanifolds. One often "normalizes" this away by looking for X as a product $Y \times Z$, and looking for submanifolds of the form

$$y \to (y, z(y)) \quad ,$$

i.e., those which are <u>graphs</u> of mappings $Y \to Z$. The "integral submanifold" equations take a simpler form: If

$$\theta = a(y,z)dy + b(y,z)dx \quad ,$$

then

$$0 = a(y,z(y))dy + b(y,z(y)) \frac{\partial z}{\partial y} = 0$$

or

$$a(y,z(y)) + b(y,z(y)) \frac{\partial z}{\partial y} = 0$$

Conversely, suppose given a set of differential equations

$$f_1\left(y, z, \frac{\partial z}{\partial y}, \ldots\right) = 0$$
$$f_2\left(y, z, \frac{\partial z}{\partial y}, \ldots\right) = 0$$
$$\ldots$$

(5.2)

One can construct exterior differential systems in terms of the variables y, z, z_y, z_{yy}, ... such that the submanifolds

$$y \to y, \quad z(y), \quad \frac{\partial z}{\partial y} = z_y, \; z_{yy} = \frac{\partial^2 z}{\partial y^2}, \ldots$$

are integral submanifolds of the exterior systems if and only if $y \to z(y)$ is a solution of the differential equations 5.2). (In fact, there are often many ways of constructing an exterior system with this property, not all necessarily equivalent.) Converting differential equations into exterior systems in this way was the key tool in the late 19-th century and early 20-th century geometric partial differential equation work that we are trying to resurrect. (See the books by Goursat and Cartan listed in the Bibliography.)

One very beautiful and useful feature of Cartan's work [4] is that he can give a precise geometric meaning (at least in case all the data is real analytic) to the classical

PROLONGATIONS

notions of "general solution", "Cauchy data", "singular solution", "characteristics", "system in involution", etc. He defines an integer called the <u>genus</u> which turns out to be the dimension of the "generically" largest dimensional integral submanifold. He also defines integers called <u>characters</u> which count the "degrees of freedom" of the "general solution". Although this material is not easy to understand or compute, its potentialities for applications in physics and engineering have hardly been touched. For example, in General Relativity, a key problem is to investigate special types of solutions of the Einstein gravitational equations. They can be set up--with a certain amount of pain--as an exterior system. (Cartan himself worked extensively on this. His papers have, to this day, not been exploited or understood <u>in detail</u>.) Further, since Cartan's methods are independent of local coordinates, it is just as easy to set up the gravitational equations in terms of "Vierbeins". (A "Vierbein" is a basis of one-forms on the four-dimensional manifold which carries the gravitational field and which are orthonormal with respect to the Riemannian metric determining the gravitational field. Cartan calls them a "moving frame", and worked extensively on their use as a computational and conceptual technique in differential geometry.)

Examples: We now revert to the usual notation of differential equation theory; independent variables are denoted by "x" or "t", dependent variables by "y" or "z". Partial derivatives--as "variables"--are denoted by subscripts: y_x, y_{xx}, etc.

a) <u>An ordinary differential equation</u>

$$f\left(x, y, \frac{dy}{dx}\right) = 0 \tag{5.3}$$

Let X be the space of variables (x, y, y_x).

$$\theta_1 = dy - y_x dx$$

$$\theta_2 = f$$

ED is generated by $\theta_1, \theta_2, d\theta_1, d\theta_2$. A one-dimensional integral submanifold restricted to $dx \neq 0$ can be (by the implicit function theorem) parameterized in the form $x \to y(x)$, and is then obviously a solution in the classical sense of the differential equation (5.2) with which we started.

b) <u>The Hamilton-Jacobi partial differential equation</u>

$$\frac{\partial S}{\partial t} = H\left(x^i, \frac{\partial S}{\partial x^i}\right) \quad , \quad i = 1 \cdots n \tag{5.4}$$

There are three ways of converting this into an ED. First, introduce variables $2n+2$, y, x^i, t, y_{x^i},

$$\theta_1 = dy - y_{x^i} dx^i - H(x^i, y_{x^i}) dt$$

PROLONGATIONS

(We use a summation convention so that $y_{x^i} dx^i = \Sigma_1^n y_{x^i} dx^i$.)
Let ED be generated by θ_1 and $d\theta_1$. We see that

$$(x^i, y) \to \left(y = S(x^i, t),\ y_{x^i} = \frac{\partial S}{\partial x^i},\ x^i,\ t \right)$$

is an n+1-dimensional integral submanifold of ED if and only if $(x^i, t) \to S(x^i, t)$ is a solution of (5.4).

However, there is another way to construct an ED whose integral submanifolds correspond to solutions of (5.4). Let X be the 2n+1-dimensional space of variables (x^i, t, y_i).

$$\theta = dy_u \wedge dx^i + dH(x^i, y_i) \wedge dt$$

$$= d(y_i dx^i + H dt)$$

Let ED be generated by the (exact!) two-form θ. On an integral submanifold of the form $(x^i, t) \to (x^i, t, y_i(x, t))$ (which is called a <u>Lagrangian submanifold</u> in current usage), i.e., on a manifold where $\phi^*\theta = 0$, so $\phi^*(y dx + H dt)$ is <u>exact</u>, there is a function $S(x^i, t)$ such that:

$$dS = \phi^*(y_i dx^i + H dt)\quad ,$$

$$\frac{\partial S}{\partial x^i} = y_i(x^i, t) \tag{5.5}$$

$$\frac{\partial S}{\partial t} = H(x^i, y_i(x^i, t))$$

$$= \quad , \text{ using } (5.5),$$

$$H\left(x^i, \frac{\partial S}{\partial x^i}\right) \quad ,$$

i.e., $S(x^i,t)$ is a solution of (5.4). The steps are obviously reversible. Thus, there is a one-one correspondence between solutions of (5.4) (when two solutions are identified if they differ by a constant) and two-dimensional integral submanifolds of ED on which

$$dx^i \wedge dt \neq 0$$

Finally, a more "space-time symmetric" version of the second approach is to "prolong", i.e., add another variable (which we denote as y_t) and a zero-form

$$\mathcal{H} = H(x^i, y_i) - y_t \quad .$$

The ED is now generated by $\mathcal{H}$, $d\mathcal{H}$ and the exact two-form

$$\theta = dy_i \wedge dx^i + dy_t \wedge dt$$

The integral manifolds of interest are those for which all $\mathcal{H} = 0$, i.e., which satisfy an <u>equation of constraint</u>. (Synge calles this "generalized homogeneous Hamiltonian mechanics [10]. See [8,9,11] for further details about using differential forms to express "constraints" in classical mechanics.)

c) The hyperbolic wave equation

$$\frac{\partial y}{\partial x \partial t} = f\left(y, \frac{\partial y}{\partial x}, \frac{\partial y}{\partial t}\right) \qquad (5.6)$$

Again, there is a standard way of writing this as an ED generated by one-forms. This involves the variables x, t, y, y_x, y_t, y_{xx}, y_{tt}. However, there is a better choice, involving fewer variables, generated by a one-form and a two-form: Introduce the variables (x,t,y,y_x,y_t).

$$\theta_1 = dy - y_x dx - y_t dt$$

$$\theta_2 = dy_x \wedge dx - f(y,y_x,y_t) dt \wedge dx \quad .$$

Let ED be generated by θ_1, θ_2, $d\theta_1$, $d\theta_2$. Again, this involves a fewer number of variables, at the expense of adding two-forms to generate the exterior system.

d) Complex function theory

Let ED be generated by the following closed two-forms in four-dimensions.

$$\theta_1 = du \wedge dx + dv \wedge dy$$

$$\theta_2 = dv \wedge dx - du \wedge dy$$

The integral manifold of the form

$$(x,y) \to (x; y, u(x,y), v(x,y))$$

are those for which

$$(x,y) \to z(x,y) = u + iv$$

is a complex analytic function, in the classical sense.

6. CONSERVATION LAWS OF DIFFERENTIAL EQUATIONS AND EXTERIOR DIFFERENTIAL SYSTEMS

We will now show how a few key concepts of differential equation theory may be described in an optimally simple and elegant form when the differential equations are first converted into exterior differential systems, as explained (briefly) in Section 5.

Let ED be an exterior differential system on a manifold X. A differential form ω is said to be a conservation law of ED if

$$d\omega \in ED \ . \tag{6.1}$$

Here is the meaning of this condition. Let

$$Y \to X$$

be an integral submanifold of X of dimension m (m-1 = degree ω) and let

$$Y_0 \to Y$$

be a submanifold of X of dimension m-1. Consider the number

PROLONGATIONS

$$\int_{Y_0} \omega \quad .$$

Suppose R is a region of Y, whose boundary ∂R has two pieces Y_0, Y_1. Here is the diagram in case Y is two-dimensional and ω is of degree one

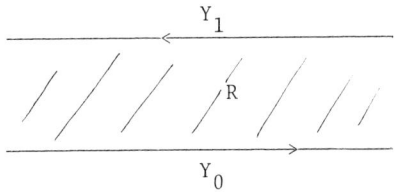

Then, because (6.1) is satisfied and because $d\omega$ is zero when pulled back to the integral submanifold Y, Stokes' formula implies that

$$0 = \int_R d\omega = \int_{Y_0} \omega - \int_{Y_1} \omega$$

This is the typical <u>conservation condition</u> arrived at when the integrand can be taken to vanish sufficiently fast toward the boundaries of Y_0.

Here are some examples:

a) <u>Conservation laws as constants of motion for ordinary differential equations</u>

Start off with a differential equation of the form

$$\frac{dx}{dt} = f(x,t) \quad . \tag{6.2}$$

Let X be the space of variables (x,t).

$$\theta = dx - fdt \ .$$

ED is generated by θ, $d\theta$.

Let h be a zero-form which is a conservation law, i.e.,

$$dh = \text{scalar times } \theta \ .$$

Now, a zero-form is a function $h(x,t)$. This condition means that

$$\frac{d}{dt} h(x(t)), t) = 0$$

when $t \to x(t)$ is a solution of (6.2) so that this coincides with the traditional idea of <u>constant of motion for ordinary differential equations</u>.

b) <u>Conservation laws for partial differential equations in two independent variables</u>

Denote the two independent variables by x,t: the dependent variable by y; its partial derivatives by $y_t, y_x, y_{xx}, \ldots$ Let X be space of variables $(x,t,y,y_x,y_t,\ldots)$ Set:

$$\theta_1 = dy - y_x dx - y_t dt$$

$$\theta_2 = dy_x - y_x dx - y_{xt} dt$$

$$\theta_3 = dy_t - y_{xt} dx - y_{tt} dt$$

etc.

PROLONGATIONS

Let ED be generated by $\theta_1, \theta_2, \theta_3, \ldots$ and a finite number of functions $f(x,t,y,y_x,\ldots)$ set equal to zero. Two dimensional integral submanifolds to which $dx \wedge dt$ restricts to be nonzero are then equivalent to functions $(x,t) \to y(x,t)$ which are solutions of the partial differential equation system

$$f(x, t, y(x,t), y_x(x,t), \ldots) = 0 \qquad (6.3)$$
$$\vdots$$

Consider now a one-differential form

$$\omega = a(x,t,y,y_x,\ldots)dx + b(x,t,y,y_x,\ldots)dt$$

To say it is a conservation law is to say that $d\omega$ is a linear combination of the two-forms on ED. Suppose it does satisfy this condition. We then see that

$$\int a(x,t,y(x,t),y_x(x,t),\ldots) \, dx$$

is independent of t, when $(x,t) \to y(x,t)$ is a solution of the partial differential equation (6.3).

7. CONSERVATION LAWS AND PROLONGATIONS

Suppose from now (until further notice) that we are dealing with a partial differential equation system of two independent variables, so that conservation laws will be one-forms. We can now use this geometric insight to generalize the notion of "conservation law" [12].

First, let us define the notion of "prolongation of an exterior differential system". Let X' and X be manifolds, n' = dimension X' $\geq$ dimension X = n

$$\pi: X' \to X$$

an <u>onto</u> map. π is called a <u>submersion map</u> if its Jacobian matrix

$$\left(\frac{\partial x'}{\partial x}\right)$$

is--in local coordinate systems for X and X' --equal to n, i.e., has "maximal rank". (In manifold theory language, this means that $\pi_*: T(X') \to T(X)$, the induced map between the tangent bundles, is <u>onto</u>.)

Suppose now that ED' is an exterior system in X', ED an exterior system on X, $\pi: X' \to X$ a submersion. π is said to define (ED') as a <u>prolongation</u> of ED if the following condition is satisfied:

$$\pi^*(\theta) \in ED'$$

$$\text{for all } \theta \in ED \ .$$

(Recall that π^* is the dual "pull-back" map.)

PROLONGATIONS

Remark. This terminology follows Lie and Cartan. In [12] the prolongation often also satisfies the following additional conditions:

> ED' is the smallest exterior differential system containing all of the $\pi^*(\theta)$
>
> The <u>genus</u> of ED' (in Cartan's sense [4]) is equal to that of ED.

Specialize this to the case where

$$X' = X \times Y ,$$

Y one-dimensional. Denote a point of X by x, a point of Y by y.

$$\pi(x,y) = x .$$

Condition (7.1) means that each $\theta \in$ ED also belongs to ED'. (Note that X' is constructed by "adding the variable y to x".) The simplest nontrivial situation then is that where:

> ED' is generated by ED, an additional one-form θ', with $d\theta' \in$ ED.

This is called a simple prolongation. θ' can then be normalized so that

$$\theta' = dy + \omega ,$$

where ω is a differential form which involves dx but not dy. We shall adopt the following notation for such differential forms: Let (x^i) be coordinates for X. ω is of the following form:

$$\omega = a_i(x,y) dx^i \quad .$$

Set:

$$\partial_y(\omega) = \left(\frac{\partial a_i}{\partial y}\right) dx^i$$

$$d_x\omega = \left(\frac{\partial a_i}{\partial x^j} dx^j\right) \wedge dx^i$$

Then,

$$d\theta' = d\omega$$

$$= -\partial_y(\omega) \wedge dy + d_x\omega$$

$$= -\partial_y(\omega) \wedge (\theta' - \omega) + d_x\omega$$

This requires that:

$$d_x\omega + \partial_y(\omega) \wedge \omega \; \varepsilon \; ED \tag{7.2}$$

for y <u>held constant</u>.

As a special case of (7.2), we have

$$d\omega \; \varepsilon \; ED \quad ,$$

i.e., ω is a conservation law if ω is independent of y.

PROLONGATIONS   37

Remark. If ω is independent of y, the variable y is
called (following [12]) a <u>potential</u>. For the general case,
i.e., where ω depends on y, y is called a <u>pseudopotential</u>.

8. MULTIPLE PSEUDOPOTENTIALS AND CARTAN-EHRESMANN
    CONNECTIONS

   Continue with the situation described in Section 7.
Let us now extend the framework by allowing the fibers of the
submersion map to have arbitrary dimensions. In fact, we
shall work locally, so that the situation can be considered
as follows.

   X, with coordinates (x^i), $1 \leq i,j \leq n$
   has an exterior system ED

   Y is a space with variables (y^a), $1 \leq a,b \leq m$

   X' = X × Y

   $\pi: X' \to X$ is the map $(x,y) \to x$

   ED' is an exterior system on X' so that π
   defines it as a prolongation of ED

<u>Definition</u>. π defines ED' as a multiple prolongation [1]
of ED if there are one-forms η^a of the form

$$\eta^a = dy^a - \omega^a , \qquad (8.1)$$

where ω^a is a linear combination of the dx^i, such that

for forms θ in ED together with the η^a generate ED'.
The fiber coordinates (y^a) of π are called (multiple)
pseudopotentials.

Notice that this type of prolongation has already been encountered in differential geometry under the name of (<u>Cartan-Ehresmann</u>) <u>connection</u> [13]. The η^a are called the <u>connection forms</u>. Thus, if Z → X is a submanifold of X which is an integral submanifold of ED, then on this submanifold the parallel transport defined by the connection is <u>independent of the path</u>. This gives a "geometric" interpretation of the whole prolongation-pseudopotential process.

Let us now work out in more detail the conditions that the forms ω^a appearing in (8.1) must satisfy:

$$d\eta^a = \partial_{y^b}(\omega^a) \wedge dy^b - d_x\omega^a$$

$$= \partial_{y^b}(\omega^a) \wedge (\eta^b + \omega^b) - d_x\omega^a \quad .$$

Thus, the conditions that ED' be <u>generated</u> by ED and the η^a is that:

$$d_x\omega^a - \omega^b \wedge \partial_{y^b}(\omega^a) \in ED \tag{8.2}$$

In the examples considered up to now [12] ω^a has the following form:

$$\omega^a = A_1^a \beta^1 + \cdots + A_r^a \beta^r \quad , \tag{8.3}$$

PROLONGATIONS 39

where $A_1^a, \ldots, A_r^a$ are functions of y alone, and $\beta^1, \ldots, \beta^r$ are one-forms in x alone. Introduce a new set of indices, $u, v, \ldots$ running from 1 to r, and the summation convention. Thus,

$$d_x \omega^a = A_u^a d\beta^u$$

$$\partial_{y^b}(\omega^a) = \partial_{y^b}(A_u^a)\beta^u$$

Hence,

$$\begin{aligned}d_x\omega^a - \omega^b \wedge \partial_{y^b}(\omega^a) &= A_u^a d\beta^u - A_v^b \beta^v \wedge \partial_{y^b}(A_u^a)\beta^u \\ &= A_u^a d\beta^u - \frac{1}{2}\left(A_v^b \partial_{y^b}(A_u^a) - A_u^b \partial_{y^b}(A_v^a)\right)\beta^v \wedge \beta^u\end{aligned} \quad (8.4)$$

The "pseudopotential" conditions are now that the <u>right hand side of</u> (8.4) <u>belongs to</u> ED. These conditions have been worked out in most detail for the Korteweg-de Vries equation [12], where they are related to a "partial" Lie algebra structure on Y. Let us now exhibit this structure.

Set:

$$A_u = A_u^a \frac{\partial}{\partial y^a} \quad (8.5)$$

Since the A_u^a were functions of y alone, this formula defines r vector fields on Y. Set:

$$A_{vu} = [A_v, A_u] \quad , \quad (8.6)$$

their Jacobi bracket. Then

$$A_{vu} = \left[A_v^b \frac{\partial}{\partial y^b}, A_u^a \frac{\partial}{\partial y^a} \right]$$

$$= A_v^b \left[\frac{\partial A_u^a}{\partial y^b} - A_u^b \frac{\partial A_v^a}{\partial y^b} \right] \frac{\partial}{\partial y^a} \quad (8.7)$$

$$= A_{vu}^a \frac{\partial}{\partial y^a}$$

Thus, we see that Equations (8.4) take the following form:

$$A_u^a d\beta^u - \frac{1}{2} A_{vu}^a \beta^v \wedge \beta^u \in ED \quad (8.8)$$

If one knew the β^u, Equation (8.8) would give relations between the A_v and A_u the $[A_u, A_v]$. In [12], the β^u are determined for the Korteweg-de Vries equation, thus leading to the following equations (with $r = 7$):

$$[A_1, A_3] = [A_2, A_3] = [A_1, A_4] = [A_2, A_6] = 0$$

$$[A_1, A_2] = -A_7 \quad ; \quad [A_1, A_7] = A_5$$

$$[A_2, A_7] = A_6 \quad ; \quad [A_1, A_5] + [A_2, A_4] = 0$$

$$[A_3, A_4] + [A_1, A_6] + A_7 = 0$$

PROLONGATIONS

One can, of course, try to "solve" these equations by <u>adding</u> additional relations to give a Lie algebra. This leads to interesting results [12], but no definite answer to its "structure" is known. However, we shall now make this further assumption and investigate the relations to the theory of Cartan-Ehresmann connections and the theory of Lie groups.

9. MULTIPLE PROLONGATIONS DEFINED BY CARTAN-EHRESMANN CONNECTIONS WITH A STRUCTURE GROUP

Let $\underset{\sim}{G}$ be a finite dimensional Lie algebra of vector fields acting in the variables y. Suppose

$$\text{dimension } \underset{\sim}{G} = r \quad,$$

and now let

$$A_u \quad, \qquad u = 1,\ldots,r \quad,$$

be a basis for $\underset{\sim}{G}$. Thus, each A_u is of the form

$$A_u = A_u^a(y) \frac{\partial}{\partial y^a} \quad,$$

where $A_u^a(y)$ are functions of y alone. There are then real numbers c_{uv}^w the <u>structure constants</u> for the Lie algebra $\underset{\sim}{G}$ with respect to this basis--such that:

$$[A_u, A_v] = c_{uv}^w A_w \quad. \tag{9.1}$$

Suppose that ED is an exterior system for X.

Now, set:

$$\eta^a = dy^a - A^a_u \beta^u . \quad (9.2)$$

Let us suppose that this formula defines the y^a as multiple pseudopotentials for ED. In differential geometric terms this means that the η^a define a connection for the fiber space $\pi: X' = X \times Y \to X$ with G (a Lie group whole Lie algebra is $\underset{\sim}{G}$) as structure group. Relation (8.5) must hold, with

$$A^a_{vu} = c^w_{vu} A^a_w , \quad (9.3)$$

i.e.,

$$A^a_u d\beta^u - \frac{1}{2} c^w_{vu} \beta^v \wedge \beta^u A^a_w \in ED . \quad (9.4)$$

Let us assume that $\underset{\sim}{G}$ acts effectively on Y, i.e., no element of $\underset{\sim}{G}$ is the zero vector field. Then relations (9.4) become the following:

$$\Omega^u = d\beta^u - \frac{1}{2} c^u_{vw} \beta^v \wedge \beta^w \in ED . \quad (9.5)$$

The two-forms Ω^u (which can be put together to define a Lie algebra--valued two-differential form on X) have a name in differential geometry--they are called the curvature forms of the connection. Of course, we obtain the equations for a conservation law in the special case where $c^u_{vw} = 0$, i.e., the Lie algebra $\underset{\sim}{G}$ is abelian. Physically, this might be interesting--it says that the theory of this type of

PROLONGATIONS

pseudopotential stands to the conservation laws as the "non-abelian" group fields, e.g., the Yang-Mills fields, stand to the abelian ones, e.g., the Maxwell field.

10. LINEAR PSEUDOPOTENTIALS AND THE INVERSE SCATTERING EQUATIONS

We have identified one type of pseudopotential-- prolongation with a Lie algebra structure acting on the fiber space Y. Clearly, different types of Lie algebras should be reflected in different sorts of pseudopotentials and prolongations. In particular, the simplest sorts of Lie algebras of vector fields--the linear ones--should, in a sense, be the simplest sort of pseudopotentials. We shall now study this situation, and show that it is, in fact, closely related to the "inverse scattering" equations associated to a nonlinear wave equation.

Let Y be a manifold with coordinates (y^a), $1 \leq a,b \leq m$. A vector field A on Y is linear if it is of the form:

$$A = \alpha_a^b y^a \frac{\partial}{\partial y^b} , \qquad (10.1)$$

with (α_a^b) constant. Vector fields of this type generate transformations on Y which are <u>linear</u> in these coordinates, since the orbit equations of A are linear, i.e.,

$$\frac{dy^b}{dt} = \alpha_a^b y^a \quad . \tag{10.2}$$

If $\underset{\sim}{G}$ is a Lie algebra of linear vector fields A on Y, then the map

$$A \to (\alpha_a^b)$$

constitutes a matrix representation of $\underset{\sim}{G}$. Conversely, a matrix representation obviously defines a linear action of $\underset{\sim}{G}$ by vector fields.

Now, let X be a space with coordinates (x^i), and with an exterior differential system ED given on X. Let Y be the space described above, with $\underset{\sim}{G}$ acting linearly. Let

$$X' = X \times Y \quad .$$

We shall look for conditions that ED' is a prolongation of ED which is generated by ED itself and a set of "pseudo-potential" forms

$$\theta^a = dy^a - A_u^a \theta^u \quad , \tag{10.3}$$

where

$$A_u = A_u^a \frac{\partial}{\partial y} \quad , \qquad 1 \leq u, v \leq r \quad ,$$

is a basis for $\underset{\sim}{G}$, and the vector fields A_u are linear, i.e., of the form

$$A_u^a = \alpha_{ub}^a y^b \quad . \tag{10.4}$$

PROLONGATIONS

The prolongation condition--leading to the generalization of "conservation law"--is that

> $d\theta^a$ lies in the Grassman algebra ideal generated by ED and θ^a. (10.5)

This implies the following property:

> If $Z \to X$, $z \to x(z)$ is an integral submanifold of ED, then the Pfaffian equations
>
> $$dy^a = \alpha^a_{ub} y^b \theta^u(x(z)) \qquad (10.6)$$
>
> are <u>completely integrable</u>.

Now the Equations (10.6) are <u>linear</u> differential equations, which are to be identified with the linear "inverse scattering" equations of nonlinear wave theory.

As an illustration, we shall now show how the Korteweg-de Vries-Schrödinger differential operators formalism (which is the prototypical example) may be viewed in this framework.

11. THE "INVERSE SCATTERING" FORMALISM AS A "COMPLETE INTEGRABILITY" CONDITION, LEADING TO THE LAX EQUATIONS. A SPECULATION ABOUT THE INVERSE SCATTERING EQUATION IN HIGHER DIMENSIONS

We shall now work more closely to the standard literature, hence will change our notation from that used in previous sections. (Recall the folk saying: Differential geometry is the study of invariants under change in notation!)

"x" denotes a real variable. We deal with linear differential operators in x which depend also on another real variable t as a parameter:

$$D = a^n(x,t)d_x^n + \cdots + a^0(x,t) \qquad (11.1)$$

($d_x = d/dx$). D_t is the operator obtained by differentiating the coefficients with respect to t.

$$D_t = a_t^n d_x^n + \cdots + a_t^0 , \qquad (11.2)$$

($d_t^n = \partial a^n/\partial t$).

Let us apply such operators to functions $\psi(x,t)$ of x and t. λ denotes a real parameter, independent of t and x.

Given a pair (D,D') of such linear operators, let us look for the condition that the following system of partial differential equations for ψ be "completely integrable".

$$D\psi = \lambda\psi \qquad (11.3)$$

$$\psi_t = D'\psi \qquad (11.4)$$

("Complete integrability" is meant in the classical sense [4].) Another term used in the classical literature is "Mayer system". In [14] it is shown that such systems are closely related to "connections". To express "complete integrability", differentiate both sides of (11.3) with respect to t, apply D to both sides of (11.4):

PROLONGATIONS   47

$$D_t\psi + D\psi_t = \lambda\psi_t \qquad (11.5)$$

$$D\psi_t = DD'\psi \qquad (11.6)$$

Insert ψ_t from (11.4) into (11.5):

$$\begin{aligned}D_t\psi + DD'\psi &= \lambda D'\psi \\ &= D'(\lambda\psi) \\ &= \text{, using (11.3)} \\ &\quad D'D\psi \qquad (11.7)\end{aligned}$$

The "rules of the game" for "complete integrability" of the system (11.3)-(11.4) now require that (11.7) hold <u>identically in</u> ψ, i.e., that

$$D_t = D'D - DD' \equiv [D',D] \quad . \qquad (11.8)$$

This, regarded as a differential equation for D, is the <u>Lax equation</u> [15] that has played the key role in the work on the inverse scattering approach to nonlinear wave soliton theory. In [16] a general Lie algebra setting for this equation is presented. It encompasses other differential equations of importance in physics--the Heisenberg equation of quantum mechanics and the Euler equation for the force-free rigid body.

We can mention here a wilder speculation. A prime question is how the formalism applies to nonlinear wave equations in more than one space variable. We believe one will set up equations like (11.3)-(11.4), then try to express the condition that they are in involution (in the classical sense [4,17]), a weaker condition than "complete integrability"

12. SPECIALIZATION TO THE KORTEWEG-DE VRIES-SCHRÖDINGER OPERATOR

Specialize the formalism described in Section 11 as follows:

$$D = d_x^2 + 2u \tag{12.1}$$

$$D' = -4d_x^3 - 6(ud_x + d_x u) \tag{12.2}$$

We obtain the Lax equation (11.8) providing u satisfies

$$u_t + 12uu_x + u_{xxx} = 0, \tag{12.3}$$

which is the Korteweg-de Vries equation. (12.1) is the Schrödinger operator, so that (11.3) is the eigenvalue problem for the Schrödinger equation of quantum mechanics (in one space variable). It is now well known how the "spectral properties" of this operator determine important and useful facts about the Korteweg-de Vries equation.

Let us now write out (11.3) and (11.4) explicitly, using (12.1)-(12.2):

$$\psi_{xx} + 2u\psi = \lambda\psi \tag{12.4}$$

PROLONGATIONS

$$\psi_t + 4\psi_{xxx} + 6(2u\psi_x + u_x\psi) = 0 \qquad (12.5)$$

Insert the value for ψ_{xx} from (12.4) into (12.5):

$$\begin{aligned} 0 &= \psi_t + 4(\lambda\psi_x - 2u_x\psi - 2u\psi_x) + 12u\psi_x + 6u_x\psi \\ &= \psi_t + 4\lambda\psi_x - 2u_x\psi + 4u\psi_x \end{aligned} \qquad (12.6)$$

We shall now interpret these equations as "linear pseudopotential" equations. Set:

$$\begin{aligned} y^1 &= \psi \\ y^2 &= \psi_x \end{aligned} \qquad (12.7)$$

(Keep in mind that superscripts "1", "2" are <u>indices</u>, not "powers". This is essential in order to consistently use the "subscript" notation for partial derivatives.) Then,

$$y^1_x = \psi_x = y^2$$

$$\begin{aligned} y^2_x &= (\psi_{xx}) \\ &= \text{, using (12.4),} \\ &\quad (-2u\psi - \lambda\psi) \\ &= (-2u - \lambda)y^1 \end{aligned}$$

This is also written in matrix form as:

$$y = \begin{pmatrix} y^1 \\ y^2 \end{pmatrix}$$

$$y_x = \begin{pmatrix} 0 & , & 1 \\ -(2u+\lambda), & & 0 \end{pmatrix} \tag{12.8}$$

Notice that this--as a linear equation--is associated with the matrix group $SL(2,R)$. (This is the tip-off that the "structure group" associated with the pseudopotentials for Korteweg-de Vries is, at least in its simplest version, $SL(2,R)$.)

Now, let us write Equations (12.5) and (12.7) as equations for the exterior derivatives dy^1, dy^2:

$$\begin{aligned} dy^1 &= y_x^1 dx + y_t^1 dt \\ &= y^2 dx + (2u_x y^1 + (4u+4\lambda)y^2)dt \end{aligned} \tag{12.9}$$

$$\begin{aligned} dy^2 &= y_x^2 dx + y_t^2 dt \\ &= -(2u+\lambda)y\,dx + (2u_{xx}y^1 + 2u_x y^2 - 4u_x y^2 \\ &\quad - (4u+4\lambda)y_x^2)dt \\ &= -(2u+\lambda)y^1 dx + (2u_{xx}y^1 - 2u_x y^2 \\ &\quad + (4u+4\lambda)(2u+\lambda)y^1)dt \end{aligned} \tag{12.10}$$

PROLONGATIONS

Equations (12.9)-(12.10) have been derived under the assumption that $u(x,t)$ is a solution of the Korteweg-de Vries equation. We know that then they are completely integrable. Let us now turn all of this around. Let X have variables

$$x,\ t,\ u,\ u_x,\ u_{xx}\ ,$$

Y have variables (y_1, y_2) and set:

$$\eta^1 = dy^1 - \omega_2^1 y^2 - \omega_1^1 y^1 \qquad (12.11)$$

$$\eta^2 = dy^2 - \omega_2^2 y^2 - \omega_1^2 y^1 \quad 1 \qquad (12.12)$$

From (12.9) and (12.10),

$$\omega_2^1 = dx - (4u + 4\lambda)dt \qquad (12.13)$$

$$\omega_1^1 = 2u_x dt \qquad (12.14)$$

$$\omega_1^2 = ((4u + 4\lambda)(2u + \lambda) + 2u_{xx})dt - (2u + \lambda)dx \qquad (12.15)$$

$$\omega_2^2 = -2u_x dt \ . \qquad (12.16)$$

We can then write (12.11)-(12.12) in matrix notation as follows

$$\eta \equiv \begin{pmatrix} \eta_1 \\ \eta_2 \end{pmatrix} = dy - \omega y \qquad (12.17)$$

(12.17) is in a good algebraic form for us to compute the complete integrability condition:

$$d\eta = -d\omega y - \omega \wedge dy$$

$$= -d\omega y - \omega \wedge (\eta + \omega y)$$

$$= (-d\omega - \omega \wedge \omega)y - \omega \wedge \eta$$

$$= -\Omega y - \omega \wedge \eta \qquad (12.18)$$

with

$$\Omega = d\omega + \omega \wedge \omega \qquad (12.19)$$

Ω is a matrix-valued differential two-form on X, called the <u>curvature form</u> of the connection. The vanishing of Ω is the integrability condition for the vanishing of η, very much the same as (11.8) is required for integrability of (11.3), (11.4). Indeed, by direct substitution, we see that Ω <u>vanishes when restricted to the submanifolds of</u> X <u>corresponding to the solutions of the Korteweg-de Vries equation</u>, i.e., Ω can be taken as generators of ED, the exterior system whose integral submanifolds are the solutions of Korteweg-de Vries. (In fact, the ED is algebraically generated by Ω, since we have the following <u>Bianchi identities</u>:

PROLONGATIONS 53

$$d\Omega = d\omega \wedge \omega - \omega \wedge d\omega$$

$$= (\Omega - \omega \wedge \omega) \wedge \omega - \omega \wedge (\Omega - \omega \wedge \omega)$$

$$= \Omega \wedge \omega - \omega \wedge \Omega \quad ,$$

which show that the exterior derivatives of Ω are already in the exterior <u>algebra</u> ideal generated by the Ω.

Remarks on "complete integrability". Standing in back of this is a general idea which might be interesting. Recall that a set $\theta_1,\ldots,\theta_n$ of one-forms is <u>completely integrable</u>, in the Frobenius sense, if $d\theta_1,\ldots,d\theta_n$ are expressible in the <u>exterior algebra</u> ideal generated by the θ. In this case, one knows completely the local structure of the integral submanifolds of the ED they generate. In fact, these integral submanifolds can be found by solving <u>ordinary</u> differential equations. In particular, they can be found in a "C^∞" context. (In general, integral submanifolds, when found by Cartan's methods [4], require "real analytic" data, since the Cauchy-Kowalewski existence theorem for systems of partial differential equations must be applied repeatedly.) Now, we see that the systems which admit solitons (and Bäcklund transformations) seem to have a property analogous to this Frobenius one--their exterior differential systems are generated by a set of <u>two</u>-forms whose exterior derivative is an exterior product of themselves with other forms. Notice that this is a priori completely disjoint from another property of Korteweg-de Vries, Sine-Gordon, etc. that some research workers call "complete integrability"--namely, the existence of an infinite number of "conservation laws" whose "Poisson brackets" are _zero_. Of course, these might well, in fact, be a relation. For example, notice in [18] how the Bäcklund transformation for Sine-Gordon "generates" the infinite hierarchy of conservation laws. An analogous relation between Bäcklund and conservation laws does _not_ seem to be satisfied for the Korteweg-de Vries, at least in as simple a form as for Sine-Gordon.

Finally, it should be noted that this method of constructing linear pseudopotential prolongations works for all those nonlinear wave equations covered by the method described in the paper, "The Inverse Scattering Transform..." by Ablowitz, Kaup, Newell, and Segur.

13. THE BÄCKLUND TRANSFORMATIONS DESCRIBED IN TERMS OF PROLONGATIONS

The Bäcklund transformations are one of the most striking and important features of the classical work on nonlinear differential equations. They had been almost forgotten until less than ten years ago, when it was recognized that they play a fundamental role in the mathematical study of solitons [18]. In this section we shall briefly describe

PROLONGATIONS

how they fit into the theory of pseudopotential prlongations; for more detail, see [14].

First, we describe the general theory, then show how it works for the Sine-Gordon equation.

Definition. Let (ED,X), (ED',X') be exterior differential systems on manifolds X and X'. Consider another pair (ED",X") consisting of an exterior differential system on a manifold X", together with a pair

$$\phi: X" \to X$$

$$\phi': X" \to X'$$

of maps. If ϕ and ϕ' define ED" as prolongations of ED and ED', i.e., if

$$\phi^*(ED) \subset ED" \supset \phi'^*(ED') \quad ,$$

then this data is said to define a Bäcklund transformation between ED and ED'.

We can see how such a Bäcklund transformation sets up a correspondence between integral submanifolds of ED and ED'. (Generally, if S_1, S_2 are sets, a correspondence between S_1 and S_2 is a subset of the Cartesian product $S_1 \times S_2$. Two elements $s_1 \in S_1$, $s_2 \in S_2$ are said to correspond if (s_1, s_2) belong to this subset. A particular kind of correspondence is obtained by taking a map $\alpha: S_1 \to S_2$, and letting the subset of $S_1 \times S_2$ be the graph of α, i.e., the set of $(s_1, \alpha(s_1))$. More general sorts of correspondences are sometimes thought of as "many-to-one maps", although such terminology is not generally accepted (for good reasons!) by the mathematical world.) Let us say that two integral submanifolds

$$Z \to X, \qquad Z' \to X'$$

correspond under the Bäcklund transformation if there is an integral submanifold

$$Z'' \to X''$$

such that:

$$\phi(Z'') = Z ; \qquad \phi'(Z'') = Z' .$$

Thus, "Bäcklund transformations" are really "pairs of prolongations". Of course, the important practical problem is to find (X'', ED''), given (X, ED), (X', ED'). When they exist, there seem to be remarkable consequences!

It can be seen that the Bäcklund transformation for the Korteweg-de Vries equation fits into this framework [12]. However, the formulas for the Bäcklund transformation are more straightforward for the Sine-Gordon equation. Accordingly, we shall now switch to this example.

PROLONGATIONS 57

$$X = \text{space of variables } (u, u_x, u_t, x, t)$$

$$\theta_1 = du - u_x dx - u_t dt \qquad (13.1)$$

$$\theta_2 = du_t \wedge dt - du_x \wedge dt - 2(\sin u) dx \wedge dt \qquad (13.2)$$

Let ED be the exterior differential system generated by θ_1, θ_2, $d\theta_1$, $d\theta_2$. The two dimensional integral submanifolds of ED on which $dx \wedge dt \neq 0$ are then the solutions of the Sine-Gordon equation:

$$u_{xt} = \sin y \qquad . \qquad (13.3)$$

Now, let X" be a space with variables

$$(x, t, u, u_x, u_t, y) \qquad .$$

Let

$$\phi: X" \to X$$

be the map

$$\phi(x, t, u, u_x, u_t, y) = (x, t, u, u_x, u_t)$$

Set:

$$\eta = dy - \left(u_x + 2\lambda \sin\left(\frac{y+u}{2}\right)\right) dx - \left(\frac{2}{\lambda} \sin\left(\frac{y-u}{2}\right) - u_t\right) dt \qquad (13.4)$$

Theorem 13.1. $d\eta$ is in the exterior algebra ideal generated by θ_1, θ_2, η. In particular, η is completely integrable on the submanifolds of (x, t, u, u_x, u_t)-space defined by solutions of the Sine-Gordon equation (14.5). The map ϕ and η define a "single pseudopotential prolongation" of the exterior system generated by θ_1, θ_2. The connection associated with this prolongation has $SL(2,R)$ as structure group

Proof.

$$d\eta = -du_x \; dx - du_t \; dt - 2\lambda \cos\left(\tfrac{y+u}{2}\right)\left(\tfrac{dy+du}{2}\right) \wedge dx$$

$$- \tfrac{2}{\lambda} \cos\left(\tfrac{y-u}{2}\right)\left(\tfrac{dy-du}{2}\right) \wedge dt$$

$$= -du_x \wedge dx + du_t \wedge dt - \lambda \cos\left(\tfrac{y+u}{2}\right) du \wedge dx$$

$$+ \tfrac{1}{\lambda} \cos\left(\tfrac{y-u}{2}\right) du \wedge dt - \lambda \cos\left(\tfrac{y+u}{2}\right)\left(\tfrac{2}{\lambda} \sin\left(\tfrac{y-u}{2}\right) - u_t\right) dt \wedge dx$$

$$- \tfrac{1}{\lambda} \cos\left(\tfrac{y-u}{2}\right)\left(2\lambda \sin\left(\tfrac{y+u}{2}\right) + u_x\right) dx \wedge dt + \cdots$$

(The terms ... involve η itself.) Thus, using the obvious trigonometric identity,

$$d\eta = -du_x \wedge dx + du_t \wedge dt - \lambda \cos\left(\tfrac{y+u}{2}\right) du \wedge dx$$

$$+ \tfrac{1}{\lambda} \cos\left(\tfrac{y+u}{2}\right) du \wedge dt + \lambda \cos\left(\tfrac{y+u}{2}\right) u_t dt \wedge dx$$

$$- \tfrac{1}{\lambda} \cos\left(\tfrac{y-u}{2}\right) u_x dx \wedge dt + 2 \sin u \; dt \wedge dx + \cdots$$

PROLONGATIONS 59

= , using (13.2),

$$\theta_2 - \lambda \cos \frac{y+u}{2} (du - u_t dt) \wedge dx + \frac{1}{\lambda} \cos \frac{y+u}{2} (du - u_x dx) \wedge dt$$
$$+ \cdots \qquad (13.5)$$

$$= \theta_2 + (\ )\theta_1 + (\ )\theta_2 + \cdots$$

This shows that η is completely integrable mod (θ_1, θ_2).

We shall now show how this result fits into the framework discussed in the previous section. Construct another space X' with variables (x', t', v, v_x, v_t). Let

$$\phi' : X'' \to X'$$

be defined via the following formula:

$$\phi'(x,t,u,u_x,u_t,y) = \left(x, t, y, \ u_x + 2\lambda \sin\left(\frac{y+u}{2}\right), \right.$$
$$\left. - u_t + \frac{2}{\lambda} \sin\left(\frac{y-u}{2}\right) \right) \qquad (13.6)$$

Set:

$$\theta_1' = dv - v_x dx' - v_t dt'$$

$$\theta_2' = dv_x \wedge dx' - 2 \sin v \, dx' \wedge dt' - dv_t \wedge dt'$$

Let ED' be the exterior differential system generated by ω_1 and ω_2. Exactly as we have seen for ED, its two-dimensional integral manifolds in which

$$dx' \wedge dt' \neq 0$$

are again a description of the solutions of the Sine-Gordon equation.

Let us compute the pull-backs under ϕ' of ED'. First,

$$\phi^*(\theta_1') = \eta \qquad (13.7)$$

and then

$$\begin{aligned}
\phi^*(\theta_2') &= d\left(u_x + 2\lambda \sin\left(\frac{y+u}{2}\right)\right) \wedge dx \\
&\quad - d\left(-u_t + \frac{2}{\lambda} \sin\frac{y-u}{2}\right) \wedge dt - 2 \sin y \, dt \wedge dx \\
&= du_x \wedge dx + du_t \wedge dt + \lambda \cos\left(\frac{y+u}{2}\right) du \wedge dx \\
&\quad + \frac{1}{\lambda} \cos\left(\frac{y-u}{2}\right) du \wedge dt + \lambda \cos\left(\frac{y+u}{2}\right) dy \wedge dx \\
&\quad - \frac{1}{\lambda} \cos\left(\frac{y-u}{2}\right) dy \wedge dt - 2 \sin y \, dt \wedge dx \\
&= du_x \wedge dx + du_t \wedge dt + \lambda \cos\left(\frac{y+u}{2}\right) du \wedge dx \\
&\quad + \frac{1}{\lambda} \cos\left(\frac{y-u}{2}\right) du \wedge dt \\
&\quad + \lambda \cos\left(\frac{y+u}{2}\right) \left(-u_t dx + \frac{2}{\lambda} \sin\left(\frac{y-u}{2}\right) dt\right) \wedge dx \\
&\quad - \frac{1}{\lambda} \cos\left(\frac{y-u}{2}\right) \left(u_x dx + 2\lambda \sin\left(\frac{y+u}{2}\right) dx\right) \wedge dt - 2 \sin y \, dt \wedge \\
&\quad + \cdots
\end{aligned}$$

PROLONGATIONS 61

$$\begin{aligned} &= -d\theta_1 + \lambda \cos\left(\frac{y+u}{2}\right) (du - u_t dt) \wedge dx \\ &\quad + \frac{1}{\lambda} \cos\left(\frac{y-u}{2}\right) (du - u_t dx) \wedge dt + \cdots \\ &= -d\theta_1 + \lambda \cos\left(\frac{y+u}{2}\right) \theta_1 \wedge dx + \frac{1}{\lambda} \cos\left(\frac{y-u}{2}\right) \theta_1 \wedge dt + \cdots \end{aligned}$$

(13.8)

We can then sum up as follows.

<u>Theorem 13.2</u>. We have the following (non-commutative) diagram of maps and exterior differential systems:

$$\begin{array}{ccc} & ED'' \quad X'' & \\ \phi \swarrow & & \searrow \phi' \\ ED, X & \xrightarrow{\alpha} & X', ED'' \end{array}$$

(The map $\alpha: X \to V$ sends (x, t, u, u_x, u_t) into (x', t', v, v_x, v_t).)

$$\alpha^*(ED'') = ED$$

$$\phi^*(ED) \subset ED''$$

$$\phi'^*(ED') \supset ED'' \quad ,$$

i.e., all maps are <u>prolongation maps</u> for the exterior systems. The pair (ϕ, ϕ') then defines a Bäcklund transformation (in the general sense described in Section 13) for the system ED whose integral manifolds are the solutions of the Sine-Gordon equation.

In summary, there is a close relation between Bäcklund transformations and prolongations of exterior differential systems. We hope that putting these ideas into this general set up (inspired, as always, by E. Cartan's method of describing properties of differential equations as "intrinsically" as possible in terms of differential forms) will lead to insights into more general systems which admit this sort of structure. We also believe that Cartan's methods are well adapted to implementation in terms of computers. Another topic of great potential interest that we have not touched on--but is obviously of great interest for applications--are the global properties of solutions of differential equations that are related via Bäcklund transformations. Since the methods we have described here are "coordinate free", they should be suitable for future research in this direction. For example, for the Sine-Gordon equation, the "bundles" constructed above are obviously the "jet" bundles associated with the space of mappings of circles into circles. Hedley Morris, in a paper in these Proceedings, has shown how the Bäcklund transformations change the "kink" (i.e., topological) structure of solutions of the Sine-Gordon equation.

BIBLIOGRAPHY

1. G. Whitham, Linear and Non-Linear Waves, Wiley, New York, 1974.
2. S. Weinberg, Gravitation and Cosmology, Wiley, N.Y., 1972.

3. R. Hermann, *Differential Geometry and the Calculus of Variations*, Academic Press, 1962 (Out of print. Second Edition under preparation via Math Sci Press).

4. E. Cartan, *Les Systèmes Exterieures et leur Applications Geométriques*, Hermann, Paris, 1946.

5. R. Hermann, "E. Cartan's Geometric Theory of Partial Differential Equations", *Advances in Math.* $\underline{1}$ (1965), pp. 265-317.

6. J. Dieudonne, *Treatise on Analysis*, Vol. IV, Academic Press, 1974.

7. W. Boothby, *An Introduction to Differentiable Manifolds and Riemannian Geometry*, Academic Press, N.Y., 1975.

8. F. Estabrook, "Some Old and New Techniques for the Practical Use of Exterior Differential Forms", in Springer Math. Lecture Notes, vol. 515, ed. R. Miura.

9. R. Hermann, *Geometric Structure of Systems-Control Theory and Physics*, Part A, Interdisciplinary Mathematics, Vol. 9, Math Sci Press, Brookline, Mass., 197 .

10. J.L. Synge, *Relativity; The General Theory*, North-Holland, Amsterdam, 1966.

11. R. Hermann, *Lie Algebras and Quantum Mechanics*, W.A. Benjamin, New York, 1970.

12. H.D. Wahlquist and F.B. Estabrook, *J. Math. Phys.* $\underline{16}$, 1975, 1-7; $\underline{17}$ (1976), 1293-1297.

13. R. Hermann, *Gauge Fields and Cartan-Ehresmann Connections*, Interdisciplinary Mathematics, Vol. 10, Math Sci Press, Brookline, Mass., 197 .

14. R. Hermann, *The Geometry of Nonlinear Differential Equations, Bäcklund Transformations, and Solitons*, Part A, Interdisciplinary Mathematics, Vol. 12, Math Sci Press, Brooline, Mass., 197 .

15. P. Lax, "Integrals of Nonlinear Equations of Evolution and Soliton Waves", *Comm. Pure Appl. Math.* $\underline{121}$, 1968, pp. 467-490.

16. R. Hermann, *Phys. Rev. Lett.* $\underline{36}$ (1976), 835; $\underline{37}$, (1976), 1591.

17. E. Goursat, Lecons sur le Problème de Pfaff, Hermann, Paris, 1922.

18. A.C. Scott, F.Y.F. Chu, and D.W. McLaughlin, "The Soliton: A New Concept in Applied Science", Proceedings of the IEEE, 61 (1973), 1443-1483.

19. R. Miura, ed., Bäcklund Transformations, Springer Math Lecture Notes, Vol. 515, 1976.

20. H. Morris, J. Math. Phys. 17 (1976), 1820-1822; 18 (1977), 533-36; 18 (1977), 530-32.

21. R.K. Dodd and J.D. Gibbon, "The Prolongation Structure of Some Higher Order Korteweg-de Vries Equations", preprint, Manchester University, 1977.

22. J. Corones, J. Math. Phys. 17 (1976), 256-57; 18 (1977), 163-164.

RECENT LITERATURE OF NEW METHODS OF DIFFERENTIAL GEOMETRY APPLIED TO NONLINEAR PARTIAL DIFFERENTIAL EQUATIONS AND SOLITON THEORY

B.K. Harrison and F.B. Estabrook, "Geometric Approach to Invariance Groups and Solution of Partial Differential Systems", J. Math. Phys. 12 (1971), 653-666.

H.D. Wahlquist and F.B. Estabrook, "Bäcklund Transformations for Solution of the Korteweg-de Vries Equation", Phys. Rev. Lett. 31 (1973), 1386-1390.

F.B. Estabrook, "Comments on Generalized Hamiltonian Dynamics", Phys. Rev. D8 (1973), 2740-2743.

F.B. Estabrook, "Some Old and New Techniques for the Practical Use of Exterior Differential Forms", in Robert M. Miura, ed., Bäcklund Transformation, the Inverse Scattering Method. Solitons and Their Application, Lecture Notes in Mathematics, No. 515, Springer-Verlag, Berlin, New York, 1976.

H.D. Wahlquist, "Bäcklund Transformation of Potentials of the Korteweg-de Vries Equation and the Interaction of Solitons with Cnoidal Waves", in R.M. Miura, ed., Bäcklund Transformation, the Inverse Scattering Method.

Solitons and Their Application, Lecture Notes in Mathematics, No. 515, Springer-Verlag, Berlin, New York, 1976.

J. Corones and F.J. Testa, "Pseudopotentials and Their Applications", in R.M. Miura, ed., Bäcklund Transformation, the Inverse Scattering Method. Solitons and Their Application, Lecture Notes in Mathematics, No. 515, Springer-Verlag, Berlin, New York, 1976.

F.B. Estabrook and H.D. Wahlquist, "The Geometric Approach to Sets of Ordinary Differential Equations and Hamiltonian Dynamics", SIAM Rev. 17 (1975), 201-220.

H.D. Wahlquist and F.B. Estabrook, "Prolongation Structures of Nonlinear Evolution Equations", J. Math. Phys. 16 (1975), 1-7.

H.C. Morris, "Prolongation Structures and a Generalized Inverse Scattering Problem", J. Math. Phys. 17 (1976), 1867-1869.

J. Corones, "Solitons and Simple Pseudopotentials", J. Math. Phys. 17 (1976), 756-759.

F.B. Estabrook and H.D. Wahlquist, "Prolongation Structures of Nonlinear Evolution Equations, II", J. Math. Phys. 17 (1976), 1293-1297.

H.C. Morris, "Prolongation Structures and Nonlinear Evolution Equations in Two Spatial Dimensions", J. Math. Phys. 17 (1976), 1870-1872.

R. Hermann, "The Pseudopotentials of Estabrook and Wahlquist, the Geometry of Solitons, and the Theory of Connections", Phys. Rev. Lett. 36 (1976), 835.

H.C. Morris, "Prolongation Structures and Nonlinear Evolution Equations in Two Spatial Dimensions, II: A Generalized Nonlinear Schrödinger Equation", J. Math. Phys. 18 (1977), 285-288.

B.K. Harrison, "Remarks on the Problem of Two Neighboring Black Holes", Utah Academy Proceedings 53 (1976), 67-74.

H.C. Morris, "A Prolongation Structure for the AKNS System and Its Generalization", J. Math. Phys. 18 (1977), 533-536.

J.C. Corones, "Solitons, Pseudopotentials and Certain Lie Algebras", J. Math. Phys. 18 (1977), 163-164.

H.C. Morris, "Prolongation Structures and Nonlinear Evolution Equations in Two Spatial Dimensions, III: A General Class of Equations", TCD 1976-7 (submitted to J. Math. Phys., 1976).

R.K. Dodd and J.D. Gibbon, "The Prolongation Structure of Some Higher Order Korteweg-de Vries Equations" (preprint, 1977).

R. Hermann, The Geometry of Nonlinear Differential Equations, Bäcklund Transformations, and Solitons, Part A, Vol. XII, Interdisciplinary Mathematics, Math Sci Press, 53 Jordan Road, Brookline, MA 02146, 1976.

H.C. Morris, "A Generalized Prolongation Structure and the Bäcklund Transformation of the Anticommuting Massive Thirring Model" TCD-1977-2 (preprint, 1977).

R. Hermann, "Quantum and Fermion Differential Geometry à ;a Cartan (preprint, 1977).

R.K. Dodd and J.D. Gibbon, "The Prolongation Structure of a Class of Nonlinear Evolution Equations" (preprint, 1977).

M. Crampin, F.A.E. Pirani and D.C. Robinson, "The Soliton Connection" (preprint, 1977).

H.C. Morris, "Prolongation Structure of Nonlinear Evolution Equations in Two and Three Dimensions", Seminar/Institute on Differential and Algebraic Geometry for Control Engineers, NASA, 1976.

H.C. Morris, "Soliton Solutions and the Higher Order Korteweg-de Vries Equations", J. Math. Phys. 18 (1977), 530-2.

R. Hermann, The Differential Geometric Foundations of the Inverse Scattering Technique and Solitons as Elementary Particles, Harvard University Physics Department preprint, 1977. To appear, Math Sci Press.

USING PSEUDOPOTENTIALS

James Corones
Iowa State University
Ames, Iowa

1. INTRODUCTION

These comments contain a brief review of pseudopotentials and several examples that illustrate the uses of pseudopotentials in the study of nonlinear partial differential equations. This is very much a report of work in progress. A great deal of progress has been made in the last two years; much of this has been in finding important structures in the pseudopotentials and in clarifying the relationship between some special, important, types of pseudopotentials and soliton equations. In the immediate future this understanding will be exploited. This paper is written with the hope that by reading a straightforward account of pseudopotentials, workers interested in soliton equations, but currently unacquainted with pseudopotentials, will be encouraged to learn more about these structures and will begin to use them.

The understanding of pseudopotentials is growing rapidly so that the assessments of the implications of pseudopotentials given in these comments must, in some sense, be considered as tentative. However, although emphasis may shift, there are three points discussed in the following whose value is sure to stay fairly fixed.

The first is that the existence of nontrivial linear pseudopotentials is closely related to the existence of (local) eigenvalue problems and isospectral flows associated with an equation. Further, it is possible in practice to distinguish classes of equations that possess nontrivial linear pseudopotentials.

The second is the existence of a Lie algebra structure in the set of pseudopotentials. This has been established in detail for equations that have simple pseudopotentials [1,2,3]. Morris has also obtained the result in [15] and has gone farther by providing an example of an interesting equation (the vector nonlinear Schrödinger equation) whose pseudopotentials have a different Lie algebra structure.

Finally, the use of pseudopotentials to find particular solutions of partial differential equations, i.e., Bäcklund transformations, is an important application. To date, these methods have only been applied to cases that can also be treated by other methods; they are, however, of a wider applicability.

These three aspects of pseudopotentials are stressed in the sequel. The reader will find other aspects of these intriguing objects discussed in the remainder of this volume.

2. A BRIEF REVIEW OF PSEUDOPOTENTIALS

In this section a brief review of pseudopotentials is given in classical notation. The classical notation is used since the primary objective of this paper is to suggest and illustrate the practical uses of pseudopotentials in the study of nonlinear partial differential equations. While the study of pseudopotentials themselves is probably best pursued in a more abstract framework, my feeling is that there is still a need to have more detailed, varied, computational results before attempting to formalize the relation between pseudopotentials and partial differential equations. For example, the Lie algebra structure of simple pseudopotentials [1,2,3] suggests an abstract study of pseudopotentials that have a Lie algebra structure [4]. However, the existence of this structure in equations that do not possess simple pseudopotentials must be established (or refuted) by example, at least at present. See, however, [5].

What is a pseudopotential? The definition is most easily given in terms of a real scalar evolution equation. So, let

$$u_t = K(u, u_x, u_{xx}, \ldots) \qquad (2.1)$$

where u is a real scalar field and K is some function of u and its space derivatives up to order $m+1$. The subscripts denote differentiation with respect to the indicated variables. Let S be the set of u and its first m space derivatives. The set of all __pseudopotentials__ associated with (2.1) is the set of all functions $q^i(x,t)$, $i = 1,\ldots,N$, N arbitrary, such that

$$q^i_x = A^i(u; q^1,\ldots,q^N; x,t) \qquad (2.2a)$$

$$q^i_t = B^i(S; q^1,\ldots,q^N; x,t) \qquad (2.2b)$$

are integrable, for all i, subject to the constraint that (2.1) is satisfied. For the connection with the forms formulation of pseudopotentials, see [1].

The integrability condition for (2.2) is:

$$A^i_t + A^i_{u_0} + \sum_{j=1}^{N} A^i_{q^j} B^j = \sum_{\ell=0}^{M} B^i_{u_0} u_{\ell+1} + \sum_{j=1}^{N} B^i_{q^j} A^j + B^i_x \qquad (2.3)$$

where $u_0 = u$, $u_1 = u_x$, $u_2 = u_{xx}$, etc., and (2.1) has been used. This condition can be written as:

$$A^i_t - B^i_x + A^i_{u_0} K - \sum_{\ell=0}^{N} B^i_u u_{\ell+1} + \sum_{j=1}^{N} \{A^i_{q^j} B^j - B^i_{q^j} A^j\} = 0 \qquad (2.4)$$

If the term in brackets vanishes for all i, the pseudopotential is called __abelian__. If this is not the case, the pseudopotential is called <u>nonabelian</u>. The extension of this definition to pseudopotentials associated with scalar equations that are not of the form (2.1) or to systems of equations is straightforward in each case. For the remainder of the discussion (except in Section 4) it will be assumed that the A^i and B^i contain no explicit space-time dependence.

It should be emphasized that the above considerations, and all that follow, are local. The original equation is local, as are (2.2). It is an interesting and open question if the notion of a pseudopotential can be extended, even in a formal sense, to either nonlocal equations or to pseudopotentials that are themselves not local.

The classical method of computing the functions A^i and B^i is discussed in [2]. The only assumption made is at the beginning of the calculation where the variable dependence of the functions is fixed, i.e., the definition of the pseudopotential. The functional form of the A^i and B^i is determined by the computation; see Section 3.

There are two topics of pseudopotentials that are of particular interest. The first is the <u>simple pseudopotential</u>, that is, the restriction of (2.2) to the case when $n = 1$. The other case is that of a <u>linear</u> pseudopotential, that is, the case when (2.2) becomes:

$$\hat{q}_x = A(u)\hat{q} \quad , \tag{2.5a}$$

$$\hat{q}_t = B(s)\hat{q} \qquad (2.5b)$$

where $\hat{q} = (q^1, q^2, \ldots, q^N)^T$ and A and B are matrix functions. Of the two cases, linear pseudopotentials are by far the most important. Simple pseudopotentials are of interest, however, for their intrinsic properties and, since they can be explicitly computed without much difficulty, for the clues they provide to general formal properties of pseudopotentials. They will therefore be discussed first.

When the simple pseudopotential for Korteweg-de Vries is computed [1], it is found that

$$q_x = \sum a_i(u) x^i(q) \qquad (2.6a)$$

$$q_t = \sum b_i(s) x^i(q) \qquad (2.6b)$$

The exact form for the a_i and b_i are of no importance for the moment. The $x^i(q)$ are found explicitly as functions of q. The bracket operation in (2.4) reduces to

$$[x^i, x^j] = x_q^i x^j - x^i x_q^j \qquad (2.7)$$

It is easily checked that this is a good Lie bracket (as is (2.4)). The x^i found for Korteweg-de Vries are closed under this bracket operation and thus form a Lie algebra. The Lie algebra structure of simple pseudopotentials is discussed in [2] and [3] and was used in a geometric theory of solitons in [4].

PSEUDOPOTENTIALS

This Lie algebra structure arises in a natural way in the simple pseudopotentials. The simple pseudopotential is explicitly computed and it is verified that the functions obtained for a Lie algebra. There is <u>no assumption</u> that the brackets are closed.

It is of interest to consider, in a brief aside, the connection between simple pseudopotentials and conservation laws. Consider the Korteweg-de Vries equation

$$u_t + uu_x + u_{xxx} = 0 \qquad (2.8)$$

This equation possess three "classical" conservation laws [1]. That is, the three pairs of functions

$$C^1 = 1 \qquad\qquad J^1 = 1$$

$$C^2 = u \qquad\qquad J^2 = -\left(\frac{1}{2} u^2 + u_{xxx}\right) \qquad (2.9)$$

$$C^3 = \frac{1}{2} u^2 \qquad\qquad J^3 = \left(\frac{1}{3} u^3 + uu_{xx} - u_x^2\right)$$

have the property that

$$C_t^i + J_x^i = 0 \qquad (2.10)$$

and that are functions of the field variables only. Clearly,

$$C = \sum_{i=1}^{3} \alpha_i C^i \quad \text{and} \quad J = \sum_{i=1}^{3} \alpha_i J^i$$

have the same property, where the α_i are constants. The potential for this conservation law is $\Phi(x,t)$ where

$$\Phi_x = \sum_{i=1}^{3} \alpha_i C^i \qquad (2.11a)$$

$$\Phi_t = - \sum_{i=1}^{3} \alpha_i J^i \qquad (2.11b)$$

The single pseudopotential for Korteweg-de Vries is obtained by letting the α_i become functions of Φ and adding a term to (2.11b), say $N(S,\Phi)$, to compensate for the functional dependence of the α_i. This function is defined by

$$[C,J+N] = N_x , \qquad (2.12)$$

where the bracket is that of (2.7). In this sense the simple pseudopotential is a nonabelian extension of the classical conservation laws. A similar argument, though more <u>ad hoc</u>, can be made for general pseudopotentials. In any case, by letting $C^i = A^i$ and $J^i = -B^i$, in (2.10) using (2.2), each pseudopotential is seen to define a conservation law, though certainly not of the classical type since they involve the q^i themselves, and not just the field variable and its derivatives. (Here "classical" is also taken to mean conservation laws that involve derivatives no higher than the order that appears in the original equation.)

Following the ideas of [6], in [1] it was shown how to use simple pseudopotentials to find Bäcklund transformations.

Given an equation, say (2.1), that possesses a simple pseudo-potential, q, the Bäcklund transformation is found in two steps. First, look for a "new" solution of (2.1), $\tilde{u}$, in terms of an "old" solution, u and q

$$\tilde{u} = \tilde{u}(u,q) \quad . \tag{2.13}$$

Again, the variable dependence of $\tilde{u}$ is assumed, but the functional form is not. This is found by substitution of (2.13) into (2.1) and use of the first order equations that q satisfies. Once $\tilde{u}$ has been found, (2.13) is solved for q in terms of u and $\tilde{u}$ and the result is substituted into the equations that define q. This yields the Bäcklund transformation. For another example of this, see [2].

The only Bäcklund transformations that have been found via the pseudopotential method up to this time, have been found by using simple pseudopotentials. The method, as outlined in [6], is probably more general, that is, would probably work for equations that possess pseudopotentials but not simple pseudopotentials; however, this should be verified in at least one instance.

The method of deriving Bäcklund transformations based on simple pseudopotentials will, obviously, only yield first order equations for the "new" solutions, that is, classical Bäcklund transformations of the Clarin-Lamb type. (See Lamb's paper in Reference [1].) Bäcklund transformations derived from more general pseudopotentials would be of a different form. This will be commented on in the concluding section.

Linear pseudopotentials are of particular importance in the study of nonlinear partial differential equations. The reason for this is as follows: if a given differential equation can have associated with it a linear eigenvalue problem and isospectral flow and if these are local, they can always be written in the form of a linear pseudopotential [1]. If no linear pseudopotentials exist for a given equation, it can have no associated eigenvalue problem and isospectral flow. When looked at in terms of linear pseudo-potentials, the problem of finding eigenvalue problems and isospectral flows is reduced to a well defined algebraic problem--that of finding solutions to certain sets of matrix equations. Thus, pseudopotentials offer a <u>computational</u> approach to finding eigenvalue problems and isospectral flows.

To make these ideas clear, in the next section the computation of linear pseudopotentials for a class of equations will be treated and the above preliminary remarks clarified in terms of this example.

3. AN EXAMPLE

Consider the equation

$$i\phi_t = -\phi_{xx} + f(\phi,\overline{\phi})\phi_x + g(\phi,\overline{\phi})\overline{\phi}_x \qquad (3.1)$$

The linear pseudopotential is given by

$$\hat{q}_x = iA(\phi,\overline{\phi})\hat{q} \qquad (3.2a)$$

$$\hat{q}_t = B(\phi,\overline{\phi},\phi_x,\overline{\phi}_x)\hat{q} \qquad (3.2b)$$

where the i in (3.2a) is introduced for convenience and A and B are $N \times N$ matrix functions of the indicated variables. Using (3.1) the integrability condition is

$$A_\phi\{-\phi_{xx} + f\phi_x + g\overline{\phi}_x\} + A_{\overline{\phi}}\{\overline{\phi}_{xx} - \overline{f}\overline{\phi}_x - \overline{g}\phi_x\} + i[A,B]$$

$$= B_\phi \phi_x + B_{\overline{\phi}}\overline{\phi}_x + B_{\phi_x}\phi_{xx} + B_{\overline{\phi}_x}\overline{\phi}_{xx} \qquad (3.3)$$

The bracket on the right hand side is the reduction of the general bracket to the linear case, that is, it is the matrix commutation of A with B. Again, subscripts denote differentiation with respect to the indicated variables.

By equating the coefficients of ϕ_{xx} and $\overline{\phi}_{xx}$, it readily follows that

$$B = \overline{\phi}_x A_{\overline{\phi}} - \phi_x A_\phi + C(\phi,\overline{\phi}) \qquad (3.4)$$

and

$$\phi_x\{fA_\phi - \bar{g}A_{\bar{\phi}}\} + \bar{\phi}_x\{gA_\phi - \bar{f}A_{\bar{\phi}}\} + i[A,B] = \phi_x B_\phi + \bar{\phi} B_{\bar{\phi}} \tag{3.5}$$

Using (3.4) in (3.5) and equating coefficients of like monomials in $\phi_x, \bar{\phi}_x$, it follows that

$$A_{\phi\phi} = 0 \tag{3.6a}$$

$$A_{\bar{\phi}\bar{\phi}} = 0 \tag{3.6b}$$

$$\{fA_\phi - \bar{g}A_{\bar{\phi}}\} - i[A,A_\phi] = C_\phi \tag{3.6c}$$

$$\{gA_\phi - \bar{f}A_{\bar{\phi}}\} + i[A,A_{\bar{\phi}}] = C_{\bar{\phi}} \tag{3.6d}$$

$$[A,C] = 0 \tag{3.6e}$$

From (3.6a,b) it follows that

$$A = \phi x_1 + \bar{\phi} x_2 + \phi\bar{\phi} x_3 + x_4 \tag{3.7}$$

where the x_i are constant matrices. Requiring that $C_{\phi\bar{\phi}} = C_{\bar{\phi}\phi}$ yields

$$\{f + \bar{f} + \bar{\phi}f_{\bar{\phi}} + \phi\bar{f}_\phi - \phi\bar{g}_{\bar{\phi}} - \bar{\phi}g_\phi\}x_3 + \{f_{\bar{\phi}} - g_\phi\}x_1 + \{\bar{f}_\phi - \bar{g}_{\bar{\phi}}\}x_2$$
$$- 2i\bar{\phi}[x_2,x_1] - 2i\phi[x_1,x_2] - 2i[x_4,x_3] \tag{3.8}$$
$$= 0$$

PSEUDOPOTENTIALS 79

Since the purpose here is to illustrate the pseudopotential method, two simple choices of f and g will be considered. In the first instance, let

$$f = \alpha\bar{\phi} \quad , \qquad g = \beta\phi \tag{3.9}$$

where α and β are real constants. It then follows from (3.8) that

$$2i[x_1, x_3] = (2\alpha - \beta)x_3 \tag{3.10a}$$

$$2i[x_2, x_3] = (2\alpha - \beta)x_3 \tag{3.10b}$$

$$2i[x_4, x_3] = (\alpha - \beta)(x_1 + x_2) \tag{3.10c}$$

If these relations are used, (3.6d,e) can be integrated to yeild

$$C = \frac{\beta}{2}\{\phi\bar{\phi}^2 - \phi^2\bar{\phi}\}x_3 + \phi\bar{\phi}P - i\phi[x_4, x_1] + i\bar{\phi}[x_4, x_3] + x_5 \tag{3.11}$$

where the notation

$$P \equiv \frac{1}{2}(\alpha + \beta)(x_1 - x_2) + i[x_1, x_2] \tag{3.12}$$

has been used. All that remains is to substitute these relations into (3.6e). Doing this and equating the coefficients of monomials in $\phi, \bar{\phi}$ to zero yields, first, the conditions

$$[x_1, x_3] = 0 \qquad (3.12a)$$

$$[x_2, x_3] = 0 \qquad (3.12b)$$

This requires, by (3.10), either $2\alpha = \beta$ or $x_3 = 0$. The second choice is taken since from the point of view of the application, the interesting solution for A in (3.2a) is that which yields an eigenvalue problem in (3.2a). Thus,

$$x_4 = \lambda \tilde{x}_4 \qquad (3.13)$$

where λ will act as the eigenvalue, is the form of x_4 of interest. Furthermore, λ cannot appear in x_1, x_2 or x_3 or in the original field equations. If $2\alpha = \beta$ the right hand side of (3.10c) must vanish if x_1, x_2, x_3 are not to depend on λ. This requires $x_1 + x_2 = 0$. In this case, x_1 commutes with x_2 and it is not difficult to show that the entire structure is abelian, i.e., all the x_i commute. Hence, $x_3 = 0$ and $\alpha = \beta$ are the appropriate choices.

Making these choices and setting $\alpha = \beta = 1$ (since the single parameter $\alpha = \beta$ can be scaled out of (3.1)) the conditions that follow from (3.6e) are

$$[x_2, (x_1-x_2) + i[x_1,x_2]] = 0 \qquad (3.14a)$$

$$[x_1, (x_1-x_2) + i[x_1,x_2]] = 0 \qquad (3.14b)$$

$$[x_1, [x_4, x_1]] = 0 \qquad (3.14c)$$

$$[x_2,[x_4,x_2]] = 0 \qquad (3.14d)$$

$$[x_4,(x_1-x_2) + 2i[x_1,x_2]] = 0 \qquad (3.14e)$$

$$[x_1,x_5] - i[x_4,[x_4,x_1]] = 0 \qquad (3.14f)$$

$$[x_2,x_5] + i[x_4,[x_4,x_2]] = 0 \qquad (3.14g)$$

$$[x_4,x_5] = 0 \qquad (3.14h)$$

It should be remarked that (3.14) admits x_4 of the form given by (3.13) by the requirement that x_5, or at least that part of x_5 that does not commute with x_1 or x_2, be proportional to λ^2. This is satisfying since, in general, (3.2b) can have rather complicated λ dependence. Indeed, in the general scheme of the inverse method, only the asymptotic $(|x| \to \infty)$ properties of the isospectral flow are needed to determine the time development of the spectral data of (3.2a).

Before discussing (3.14) further, a second example, an example of the usual situation, will be discussed. To do this, return to (3.8) and now let

$$f = (\phi\bar{\phi})^5 \, , \qquad g = 0 \qquad (3.15)$$

In this case (3.8) implies

$$x_1 = x_2 = x_3 = 0 \, , \qquad (3.16)$$

and there is therefore no nontrivial linear pseudopotential. Hence no local eigenvalue problem and isospectral flow exist for (3.1) with f and g given by (3.15). The equation so defined is not of any particular interest in itself, however, it is the sharpness of the result that is to be emphasized.

It is not difficult to see that by this method, classes of equations could be treated, say,

$$f = \alpha \phi^p \bar{\phi}^r \qquad g = \beta \phi^s \bar{\phi}^t \qquad (3.17)$$

If sufficient care is taken, it is possible to derive conditions on p, r, s, t, α and β that must be satisfied if any nontrivial linear pseudopotential is to exist. At the same time, the explicit field variable dependence of A and B is derived together with algebraic conditions analogous to (3.19). The calculation thus finds all possible equations in the given class which can have local eigenvalue problems and isospectral flows and reduces the problem of finding them to a well defined algebraic problem.

The question of solving (3.14) or their analogue at present goes beyond what can be learned by the study of pseudopotentials per se. However, the existence of the Lie algebra structure in linear pseudopotentials [3,5] may give some useful clues to solving these structure equations. It is not yet clear that the existence of the Lie algebra structure is necessary, though a result of this kind is certainly anticipated. One point is clear though: the parameter that plays the role of the eigenvalue in (3.2a) should not appear in the structure constants of the Lie algebra. This happens when the structure is closed "by hand" as in [6] and [7]. In fact, it appears that the eigenvalue should be factored out of the structure equations (see the argument after (3.13)). Only by doing this is the common algebraic structure of equations solvable by the generalized Zakharov-Shabat problem [3,5] evident.

At present it is difficult to see the advantage of computing the most general pseudopotentials associated with an equation and then search for a linear representation. Or, said another way, what can be learned for nonlinear pseudopotentials that cannot be learned from linear pseudopotentials? If a Lie algebra structure is found, what is the advantage of using nonlinear rather than linear representation?

4. SPACE-TIME DEPENDENT PSEUDOPOTENTIALS

It is natural to assume that the pseudopotentials associated with a given equation do not explicitly depend on space and time if the equation itself has constant coefficients. If variable coefficient equations are to be studied, this assumption should be dropped. The question arises as to whether there exist variable coefficient equations that can be solved by the inverse scattering method, using eigenvalue problems and isospectral flows that are constrained by the method outlined above.

David Levermore and I are examining this question using the variable coefficient nonlinear Schrodinger equation

$$i\phi_t + \alpha\phi + \beta\phi_x + \gamma\phi_{xx} + \delta|\phi|^2\phi = g \quad (4.1)$$

as a starting point. In our analysis α, β, γ, δ and g are all taken to be functions of space and time.

The first step is to find the linear space-time dependent pseudopotentials associated with (4.1). That is, to find A and B such that

$$\hat{q}_x = iA(\phi,\overline{\phi},x,t)\hat{q} \quad (4.2a)$$

$$\hat{q}_t = B(\phi,\overline{\phi},\phi_x,\overline{\phi}_x,x,t)\hat{q} \quad (4.2b)$$

The following results are obtained

$$A = \phi Z_1 + \overline{\phi} Z_2 + Z_4 \quad (4.3a)$$

$$B = -\gamma\phi_x Z_1 + \overline{\gamma\phi}_x Z_2 + i\gamma\phi\overline{\phi}[Z_1,Z_2] + \phi Q + \overline{\phi} R + Z_5 \quad (4.3b)$$

where

$$Q = -\beta Z_1 - i\gamma[Z_4,Z_1] + (\gamma Z_1)_x \quad (4.4a)$$

$$R = \overline{\beta} Z_2 + i\gamma[Z_4,Z_2] - (\overline{\gamma} Z_2)_x \quad (4.4b)$$

where the Z_i are matrix functions of space and time and the subscripts again denote differentiation. The Z_i are constrained by the following differential equations

$$-\delta Z_1 - \gamma[Z_1,[Z_1,Z_2]] = 0 \qquad (4.5a)$$

$$\bar{\delta} Z_2 - \gamma[Z_2,[Z_1,Z_2]] = 0 \qquad (4.5b)$$

$$-\gamma[Z_4,[Z_1,Z_2]] + i[Z_1,R] + i[Z_2,Q] = i(\gamma[Z_1,Z_2])_x \qquad (4.5c)$$

$$[Z_1,Q] = 0 \qquad (4.5d)$$

$$[Z_2,R] = 0 \qquad (4.5e)$$

$$-\alpha Z_1 + i[Z_1,Z_5] + i[Z_4,Q] + Z_{1t} = Q_x \qquad (4.5f)$$

$$\bar{\alpha} Z_2 + i[Z_2,Z_5] + i[Z_4,R] + Z_{2t} = R_x \qquad (4.5g)$$

$$g Z_1 - \bar{g} Z_2 + i[Z_4,Z_5] + Z_{4t} = Z_{5x} \qquad (4.5h)$$

Since it was by no means obvious that any variable coefficient equations can be solved by the inverse method, a simple case of (4.1) and, hence (4.5), was first studied. The case $\alpha = \beta = g = 0$, γ constant, was treated. Just as in the case of constant coefficient equations, where the requirement of a nonabelian pseudopotential restricts the general form of the equation, in the variable coefficient case the form of the coefficients is restricted by the same requirement.

In the present case, two differential equations are found for $\delta(x,t)$, the solution of which is

$$\delta = \frac{k_0}{1 + \beta t} \qquad (4.6)$$

where k_0 and β are real constants. In addition, (4.2a) yields an eigenvalue problem where the role of the eigenvalue is played by the function $\lambda/(1+\beta t)$ where λ is constant. However, the inverse scattering method can be generalized (slightly) to cover this case. Indeed, explicit solutions of the equations can be constructed by this method. The details of this and additional work on (4.1) will be reported elsewhere.

From this it appears that space-time dependent pseudopotentials provide a method of finding eigenvalue problems and isospectral flows for variable coefficient equations. The method can be thought of as a sort of variation of parameters approach. Since it is known that the generators of $SU(2)$ form a basis for the matrix solutions of (4.5) when (4.1) has constant coefficients [3], the most direct method of solving (4.5) is to assume that the Z_i are combinations of these generators with variable coefficients. An equivalent approach is to find the space-time dependent simple pseudopotential for (4.1).

BIBLIOGRAPHY

1. J. Corones and F. Testa, in <u>Backlund Transformations</u>, Proceedings of NFS Workshop, <u>Vanderbilt University</u>, September 1974. Spring Lecture Notes in Mathematics No. 15 (1976).

2. J. Corones, <u>J. Math. Phys</u>. <u>17</u>, 756 (1976).

3. J. Corones, "Solitons, pseudopotentials and certain Lie algebras", to appear in <u>J. Math. Phys</u>. Jan.1977.

4. R. Hermann, <u>Phys. Rev. Lett</u>. <u>36</u>, 835 (1976).

5. H. Morris, "Prolongation structure for the AKNS system and its generàlizations", preprint.

6. H. Wahlquist and F. Estabrook, <u>J. Math. Phys</u>. <u>16</u>, 1 (1975).

7. H. Morris, "Prolongation structures and nonlinear evolution equations in two spatial dimensions", preprint.

DIFFERENTIAL GEOMETRIC VIEWPOINTS ON THE DEVELOPMENT OF SHOCK WAVES

Robert B. Gardner
University of North Carolina
Chapel Hill, North Carolina

This report will describe a method for solving initial value problems by determining systems of ordinary differential equations whose solutions extend the initial data to a solution of the original problem. After giving the construction in the simplest setting, the method will be applied in detail to the non-linear hyperbolic equations in the plane.

Let $I = \{\omega^1, \ldots, \omega^p\}$ be a Pfaffian system, that is, a system of total differential equations, or 1-forms, on some Euclidean space $\mathbb{R}^N$. Let

$$\Delta: \mathbb{R}^k \to \mathbb{R}^N$$

be Cauchy data, that is, a one-one differentiable map with $\Delta^* I = 0$. Equivalently, Δ defines a k-dimensional solution of I. Now every solution of the Cauchy problem of extending Δ to a (k+1)-dimensional solution of I, that is, every map

$$\sigma: \mathbb{R}^k \times \mathbb{R} \to \mathbb{R}^N$$

satisfying

$$\sigma^* I = 0 \quad \text{and} \quad \sigma(u,0) = \Delta(u)$$

for $u \in \mathbb{R}^k$

can be realized as the extension of the initial data by the integral curves of a vector field of system of ordinary differential equations. That is,

$$\sigma(u,t) = \exp_{\Delta(u)} tX$$

where X is a vector field and the symbol on the right denotes the flow at time t which started at $\Delta(u)$ at time zero.

We will call a vector field X k-stable for I if

1) $X \in I^{\perp}$

and

2) The sequence of systems of 1-forms

$$D_X^0(I) = I, \ldots, D_X^j(I)$$
$$= \{D_X^{j-1}(I), L_X D_X^{j-1}(I)\}, \ldots$$

stabilizes with

$$D_X^k(I) = D_X^{k+1}(I) \ .$$

Note that 0-stable is the condition of the classical Cauchy characteristics

$x \in I^\perp$ and $L_X I \subset I$.

With this preparation we can now state the basic result (see [1]).

Theorem 1. The Cauchy problem for I with Cauchy data Δ has a solution

$$\sigma(u,t) = \exp_{\Delta(u)} tX$$

for a k-stable X if and only if

1) X is transversal to image Δ
2) $\Delta^* D_X^k(I) = 0$.

There is a class of systems I for which this method is especially amenable. We will say that I has characteristics for q-dimensional solutions if

1) There exists vector fields X called characteristic vector fields such that

    $\dim D_X^{(1)}(I)$ is minimal and less than the dimension for general X.

2) Every q-dimensional solution of I is tangent to a characteristic vector field.

In case I has characteristics the question of existence of solutions is reduced to the study of k-stability of only the characteristic vector fields.

The rest of this report will be devoted to showing how non-linear hyperbolic equations in the plane may be solved by applying the above ideas to an associated system of 1-forms which admit characteristics.

A second order partial differential equation

$$F(x,y,z,z_x,z_y,z_{xx},z_{xy},z_{yy}) = 0 \tag{1}$$

can be viewed as a 7-dimensional submanifold of the eight dimensional space of equivalence classes of functions having the same value at a point and the same derivatives up through order two at the point. The equivalence class of a function

$$f: \mathbb{R}^2 \to \mathbb{R}$$

at $(x,y) \in \mathbb{R}^2$ will be denoted by $j^2_{(x,y)}(f)$. We introduce coordinates on the set of equivalence classes, denoted by $J^2(\mathbb{R}^2, \mathbb{R})$, by

$$x(j^2_{(x,y)}(f)) = x \quad,$$

$$y(j^2_{(x,y)}(f)) = y \quad,$$

$$z(j^2_{(x,y)}(f)) = f(x,y) \quad,$$

$$p(j^2_{(x,y)}(f)) = \frac{\partial f}{\partial x}(x,y) \quad,$$

$$q(j^2_{(x,y)}(f)) = \frac{\partial f}{\partial y}(x,y) \quad,$$

SHOCK WAVES

$$r(j^2_{(x,y)}(f)) = \frac{\partial^2 f}{\partial x^2}(x,y) \quad ,$$

$$s(j^2_{(x,y)}(f)) = \frac{\partial^2 f}{\partial x \partial y}(x,y) \quad ,$$

$$t(j^2_{(x,y)}(f)) = \frac{\partial^2 f}{\partial y^2}(x,y) \quad .$$

As such, the problem of solving (1) is the problem of finding a function f with $j^2_{(x,y)}(f)$ contained in the locus,

$$F(x,y,z,p,q,r,s,t) = 0 \quad .$$

Since the mappings of the form $\gamma(x,y) = j^2_{(x,y)}(f)$, that is the 2-graphs, are not the general mappings into the space of equivalence classes, it is natural to classify those which are 2-graphs as functions. If

$$\gamma(x,y) = (x,y,z(x,y),p(z,y),q(x,y),r(x,y),s(x,y),t(x,y))$$

is a 2-graph, then we have relations

$$\frac{\partial z}{\partial x} = p \quad , \quad \frac{\partial z}{\partial y} = q \quad ,$$

$$\frac{\partial^2 z}{\partial x^2} = r \quad , \quad \frac{\partial^2 z}{\partial x \partial y} = s \quad , \quad \frac{\partial^2 z}{\partial y^2} = t$$

which can be written more succintly in the form

$$\gamma^* \begin{cases} \tilde{\omega}^1 = dz - pdx - qdy \\ \tilde{\omega}^2 = dp - rdx - sdy \\ \tilde{\omega}^3 = dq - sdx - tdy \end{cases} = 0 \qquad (2)$$

Conversely, it follows that a mapping $\gamma(x,y)$ which satisfies (2) is the 2-graph of a function.

Now let

$$\Sigma_7 \xrightarrow{i} J^2(\mathbb{R}^2, \mathbb{R})$$

be a 7-manifold whose image under i gives the locus

$$F(x,y,z,p,q,r,s,t) = 0 \ .$$

The problem of solving the partial differential equation is now equivalent to finding a mapping $\tilde{\sigma}$

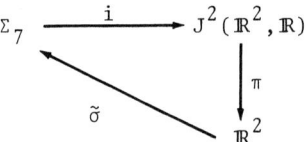

where $i \circ \sigma(x,y) = j^2_{(x,y)}(f)$, or, by what was just discussed, a mapping

$$\tilde{\sigma}: \mathbb{R}^2 \to \Sigma_7$$

satisfying

$$\pi \circ i \circ \tilde{\sigma} = 1d \ ,$$

and

$$\tilde{\sigma}^* \begin{cases} \omega^1 = i^*\tilde{\omega}^1 \\ \omega^2 = i^*\tilde{\omega}^2 = 0 \\ \omega^3 = i^*\tilde{\omega}^3 \end{cases}$$

The first condition is a bit clumsy in practice, but can be replaced by the more natural problem of finding a mapping

$$\sigma: \mathbb{R}^2 \to \Sigma_7$$

satisfying

$$\pi \circ i \circ \sigma \text{ is a diffeomorphism}$$

and

$$\sigma^* \begin{cases} \omega^1 \\ \omega^2 \\ \omega^3 \end{cases} = 0$$

This is seen by noting that given such a σ we construct a $\tilde{\sigma}$ by

$$\tilde{\sigma} = i \circ \sigma \circ (\pi \circ i \circ \sigma)^{-1} .$$

Thus we have arrived at the position that solving a second order partial differential equation is equivalent to finding two-dimensional integral manifolds of a three-dimensional Pfaffian system on a 7-manifold Σ_7.

Let us view $I = \{\omega^1, \omega^2, \omega^3\}$ as a module over the C^∞-functions on Σ_7. Given $\phi, \psi \in I$ we introduce an intrinsic conformal symmetric bilinear form on I by letting $<\phi,\psi>$ be defined by

$$d\phi \wedge d\psi \wedge \omega^1 \wedge \omega^2 \wedge \omega^3 = <\phi,\psi> d\Sigma_7$$

where $d\Sigma_7$ is any volume form on Σ_7. The linearity over the functions follows from the observation that

$$d(f\psi) \wedge \omega^1 \wedge \omega^2 \wedge \omega^3 = df \wedge \psi \wedge \omega^1 \wedge \omega^2 \wedge \omega^3 + fd\psi \wedge \omega^1 \wedge \omega^2 \wedge \omega^3$$

and

$$\psi \wedge \omega^1 \wedge \omega^2 \wedge \omega^3 = 0$$

since ψ is linearly dependent on $\{\omega^1, \omega^2, \omega^3\}$.

In order to relate the bilinear form with the defining equation we note that an equivalent definition of the bilinear form is the pullback of the form defined on $\tilde{I} = \{\tilde{\omega}^1, \tilde{\omega}^2, \tilde{\omega}^3\}$ by defining $<\tilde{\phi}, \tilde{\psi}>$ for $\tilde{\phi}, \tilde{\psi} \in \tilde{I}$ by

$$d\tilde{\phi} \wedge d\tilde{\psi} \wedge \tilde{\omega}^1 \wedge \tilde{\omega}^2 \wedge \tilde{\omega}^3 \wedge dF = <\tilde{\phi}, \tilde{\psi}> d\Sigma_8$$

where $d\Sigma_8$ is any volume form on $J^2(\mathbb{R}^2, \mathbb{R})$.

An easy computation then gives that the matrix of $<,>$ relative to the basis $\{\omega^1, \omega^2, \omega^3\}$ has representation

$$\begin{pmatrix} 0 & 0 & 0 \\ 0 & F_t & -\tfrac{1}{2}F_s \\ 0 & -\tfrac{1}{2}F_s & F_r \end{pmatrix} \tag{3}$$

SHOCK WAVES

The symmetric matrices of rank two are classified by the sign of their determinant and the sign of

$$\Delta = F_t F_r - \frac{1}{4} F_s^2$$

recovers the classical partitioning of points on partial differential equation by type. Note that by construction, the type is invariant under contact transformations, that is, under diffeomorphism

$$f: J^2(\mathbb{R}^2, \mathbb{R}) \to J^2(\mathbb{R}^2, \mathbb{R})$$

satisfying

$$f^* I = I \quad .$$

The rest of our results will be based on the assumption that the equation (1) is hyperbolic, that is,

$$\Delta = F_t F_r - \frac{1}{4} F_s^2 < 0 \quad .$$

In this case, there is a change of basis for I to $\{\omega^1, \pi^2, \pi^3\}$ relative to which the matrix of $\langle \, , \rangle$ has representation

$$\begin{pmatrix} 0 & 0 & 0 \\ 0 & 0 & 1 \\ 0 & 1 & 0 \end{pmatrix} \tag{4}$$

This implies that there exist linearly independent forms $\{\omega^1, \pi^2, \pi^3, \omega^4, \omega^5, \omega^6, \omega^7\}$ such that

$$d\pi^1 \equiv 0$$
$$d\pi^2 \equiv \omega^4 \wedge \omega^5 \quad \text{mod I} \qquad (5)$$
$$d\pi^3 \equiv \omega^6 \wedge \omega^7$$

Example. (Wave equation) $r - t = 0$. The matrix (3) is given by

$$\begin{pmatrix} 0 & 0 & 0 \\ 0 & -1 & 0 \\ 0 & 0 & 1 \end{pmatrix}$$

which may be put into the form (4) by choosing

$$\pi^2 = \omega^3 - \omega^2 \quad \text{and} \quad \pi^3 = \omega^3 + \omega^2 \quad .$$

Then

$$d\pi^1 \equiv 0$$
$$d\pi^2 \equiv d(x-y) \wedge d(s-t) \quad \text{mod I}$$
$$d\pi^3 \equiv d(x+y) \wedge d(s+t)$$

Hence we may choose

$$\omega^4 = d(x-y) , \quad \omega^5 = d(s-t) , \quad \omega^6 = d(x+y) ,$$
$$\omega^7 = d(s+t) \quad .$$

Note it is not generally true that $\omega^4, \omega^5, \omega^6, \omega^7$ may be chosen to be exact.

SHOCK WAVES

Now we proceed to apply the method described in the beginning of the report. The first step is to study the systems $D_X^{(1)}(I)$ for $X \in I^\perp$ and determine whether characteristic vector fields exist.

Since $D_X^{(1)}(I)$ will be minimal when the kernel of L_X is maximal, we must compute the Lie derivatives of linear combinations of I.

Since $X \in I^\perp$ and $L_X = X \,\lrcorner\, d + dX \,\lrcorner\,$, Equations (5) give

$$L_X \omega^1 \equiv 0 \mod I$$

and

$$L_X(\alpha \pi^2 + \beta \pi^3) \equiv \alpha X \,\lrcorner\, d\pi^2 + \beta X \,\lrcorner\, d\pi^3$$

$$\equiv \alpha \langle X, \omega^4 \rangle \omega^5 - \alpha \langle X, \omega^5 \rangle \omega^4$$

$$+ \beta \langle X, \omega^6 \rangle \omega^7 - \beta \langle X, \omega^7 \rangle \omega^6 \mod I$$

As a result, the condition

$$L_X(\alpha \pi^2 + \beta \pi^3) \equiv 0 \mod I$$

leads to three cases:

I) $\alpha \neq 0$, $\beta \neq 0$, whence $X = 0$

II) $\alpha = 0$, $\beta \neq 0$, whence $X \in \{I, \omega^6, \omega^7\}^\perp = V_1$

III) $\alpha \neq 0$, $\beta = 0$, whence $X \in \{I, \omega^4, \omega^5\}^\perp = V_2$.

Thus,

$$\text{minimal dim } D_X^{(1)}(I) = 4 ,$$

which is less than the generic dimension, which is five, if and only if X lies in one of the two two-dimensional vector field systems V_1 or V_2.

Next we must check to see whether every solution is swept out by vector fields from these vector field systems.

Let $\sigma: \mathbb{R}^2 \to \Sigma_7$ be a solution, then $\sigma^* I = 0$ and Equation (5) yield

$$\sigma^*(\omega^4 \wedge \omega^5) = 0 \quad \text{and} \quad \sigma^*(\omega^6 \wedge \omega^7) = 0 \ .$$

Now since a solution must have Jacobian of rank two,

$$\dim\{\sigma^* V_1^\perp\} = 1 \quad \text{and} \quad \dim\{\sigma^* V_2^\perp\} = 1 \ .$$

Now let

$$X_1 \in \{\sigma^* V_1^\perp\}^\perp \quad \text{and} \quad X_2 \in \{\sigma^* V_2^\perp\}^\perp$$

then

$$\sigma_* X_1 \in V_1 \quad \text{and} \quad \sigma_* X_2 \in V_2$$

This proves that every solution is tangent to a vector field in V_1 and V_2 and proves that I admits characteristics for two-dimensional solutions.

As such, the initial value problem depends entirely on the analysis of the K-stability condition for the characteristic vector fields.

Let us consider $X \in V_1 = \{I, \omega^4, \omega^5\}^\perp$ then

$$D_X^{(1)}(I) = \{I, X \rfloor d\pi^3\} \ ,$$

SHOCK WAVES

and by (5)

$$X \lrcorner\, d\pi^3 \equiv \langle X,\omega^6\rangle \omega^7 - \langle X,\omega^7\rangle \omega^6 \qquad \text{mod } I \ .$$

Now, if $X \neq 0$, then not both $\langle X,\omega^6\rangle$ and $\langle X,\omega^7\rangle$ are zero. By possibly making the modification

$$\bar\omega^6 = \omega^7 \ , \qquad \bar\omega^7 = -\omega^6$$

we may assume $\langle x,\omega^6\rangle \neq 0$ and deduce that

$$D_X^{(1)}(I) = \{I, \omega^7 - \lambda\omega^6\}$$

where λ is the function $\langle X,\omega^7\rangle / \langle X,\omega^6\rangle$. The next system in the sequence is

$$D_X^{(1)}(I) = \{I, \omega^7 - \lambda\omega^6,\ X \lrcorner\, d(\omega^7 - \lambda\omega^6)\}$$

and the condition of 1-stability for X becomes

$$X \lrcorner\, d(\omega^7 - \lambda\omega^6) \wedge (\omega^7 - \lambda\omega^6) \equiv 0 \qquad \text{mod } I \qquad (6)$$

This last equation holds if and only if the wedge of the left side with ω^4, ω^5, and ω^6 are congruent to zero modulo I, and these equations are equivalent to the wedge of the left side of (6) with $\omega^4 \wedge \omega^5$ and ω^6 being congruent to zero modulo I. Now, since

$$\{X\}^{\perp} = \{I, \omega^4, \omega^5, \omega^7 - \lambda\omega^6\} \ ,$$

and

$$d(\omega^7 - \lambda\omega^6) \wedge (\omega^7 - \lambda\omega^6) \wedge \omega^4 \wedge \omega^5 \wedge \omega^1 \wedge \pi^2 \wedge \pi^3 = 0 \quad (7)$$

being an eight form on a 7-manifold, we see

$$X \lrcorner d(\omega^7 - \lambda\omega^6) \wedge (\omega^7 - \lambda\omega^6) \wedge \omega^4 \wedge \omega^5 \equiv 0 \mod I .$$

As a result, the condition of 1-stability for X is equivalent to

$$X \lrcorner d(\omega^7 - \lambda\omega^6) \wedge (\omega^7 - \lambda\omega^6) \wedge \omega^6 \equiv 0 \mod I$$

or

$$\langle X, \omega^6 \rangle d\lambda \equiv - X \lrcorner d\omega^7 + \lambda X \lrcorner d\omega^6 \equiv 0 \mod(I, \omega^6, \omega^7) \tag{8}$$

which is a system of two quasi-linear first order partial differential equations for λ.

Example. (Fermi-Pasta-Ulam equation) $t - k^2(p)r = 0$. By Equation (3) the matrix of $\langle , \rangle$ is

$$\begin{pmatrix} 0 & 0 & 0 \\ 0 & 1 & 0 \\ 0 & 0 & -k^2 \end{pmatrix}$$

hence we choose

$$\pi^2 = k\omega^2 + \omega^3, \qquad \pi^3 = k\omega^2 - \omega^3$$

Now,

$$d\pi^2 \equiv kd\omega^2 + d\omega^3$$

$$\equiv - k(dr \wedge dx + ds \wedge dy) - (ds - 2kk'r^2 dy) \wedge dx - k^2 dr \wedge dy$$

$$\equiv (dx + kdy) \wedge (ds + kdr - 2kk'r^2 dy) .$$

SHOCK WAVES

Hence we choose

$$\omega^4 = dx + kdy, \qquad \omega^5 = ds + kdr - 2kk'r^2 dy$$

and similarly

$$d\pi^3 \equiv (dx - kdy) \wedge (kdr - ds + 2kk'r^2 dy)$$

and we choose

$$\omega^6 = dx - kdy, \qquad \omega^7 = kdr - ds + 2kk'r^2 dy .$$

The two families of characteristics are now

$$V_1 = \{I, \omega^4, \omega^5\}^\perp \quad \text{and} \quad V_2 = \{I, \omega^6, \omega^7\}^\perp .$$

Let us restrict our attention to the first family V_1 which is easily determined to be

$$V_1 = \left\{ -k\tilde{\partial}_x + \tilde{\partial}_y + 2kk'r^2 \frac{\partial}{\partial s} - \lambda \left(\frac{\partial}{\partial r} - k \frac{\partial}{\partial s} \right) \right\}$$

where

$$\tilde{\partial}_x = \frac{\partial}{\partial x} + p \frac{\partial}{\partial z} + r \frac{\partial}{\partial p} + s \frac{\partial}{\partial q}$$

and

$$\tilde{\partial}_y = \frac{\partial}{\partial y} + q \frac{\partial}{\partial t} + s \frac{\partial}{\partial p} + t \frac{\partial}{\partial q}$$

and λ as before is the function $<x, \omega^7>/<x, \omega^6>$.

In particular, we see that up to a factor only affecting speed, the existence of a 1-stable characteristic field $X \in V_1$ is completely determined by the existence of the function λ satisfying (8). These Equations (8) imply

$$\lambda = \underbrace{(K's/2K)r - 5/4\ K'r^2}_{\text{quadratic in } r} - Kr + f(x,y,z,p,q,s)$$

and the usual comparison theorm (see Theorem 1 in [2]) implies finite escape time in the r-variable and hence a second order shock in that derivative.

BIBLIOGRAPHY

1. Gardner, R., "The Cauchy Problem for Pfaffian Systems", Comm. Pure and Appl. Math. 22 (1969), 587-596.

2. Lax, P., "Development of Singularities of Solutions of Non-Linear Hyperbolic Partial Differential Equations", J. Math. Phys. 5 (1964), 611-613.

BÄCKLUND TRANSFORMATIONS AND
THE SINE-GORDON EQUATION

Hedley C. Morris
Trinity College
Dublin, Ireland

1. PROLONGATIONS

Suppose that we have a manifold M and an ideal of differential n-forms I. An $F(M)$-submodule P of the set of (n-1) forms on M is defined to be a <u>prolongation</u> of I if

$$dP \subset F^*(M) \wedge P + I \quad .$$

All of the equations that have so far been solved by the inverse scattering method possess a prolongation structure. The ideal I spanned by the three two-forms

$$\alpha_1 = dA \wedge dt + (rB - qC)dx \wedge dt \quad (1.1)$$

$$\alpha_2 = dB \wedge dt + dq \wedge dx + 2[i\lambda B + Aq]dx \wedge dt \quad (1.2)$$

$$\alpha_3 = dC \wedge dt + dr \wedge dx - 2[i\lambda C + Ar]dx \wedge dt \quad (1.3)$$

is a closed ideal as

$$d\alpha_1 = \alpha_3 \wedge (Bdt + qdx) - \alpha_2 \wedge (Cdt + rdx) \qquad (1.4)$$

$$d\alpha_2 = 2[\alpha_2 \wedge (Adt - i\lambda dx) - \alpha_1 \wedge (qdx)] \qquad (1.5)$$

$$d\alpha_3 = 2[\alpha_3 \wedge (-Adt + i\lambda dx) + \alpha_1 \wedge (rdx)] \qquad (1.6)$$

and consequently, by Cartan's theory [1], is completely equivalent to the AKNS equations [2]

$$A_x = qC - rB \qquad (1.7)$$

$$q_t = B_x + 2(Aq + i\lambda B) \qquad (1.8)$$

$$r_t = C_x - 2(Ar + i\lambda C) \qquad (1.9)$$

which give rise to most of the equations solvable by inverse scattering methods. Using the methods of Wahlquist and Estabrook [3] it can be shown [4] that the one-form

$$\Omega = dy - [r + 2i\lambda y - qy^2]dx - [C - 2Ay - By^2]dt \qquad (1.10)$$

gives the prolongation of I. In fact,

$$d\Omega = 2y\alpha_1 + y^2\alpha_2 - \alpha_3 + [(i\lambda - qy)dx - (A + By)dt] \; 2\Omega \qquad (1.11)$$

The new variable y is known as a <u>Pseudopotential</u> for I. In general, there will be several pseudopotentials. For example, the Bousinesq equation

$$\frac{3}{4} u_{tt} + \frac{1}{4} u_{xxxx} + \frac{3}{2} (uu_x)_x = 0 \qquad (1.12)$$

BÄCKLUND TRANSFORMATIONS

can be represented by the ideal spanned by the four two-forms

$$\alpha_1 = du \wedge dt - p\, dx \wedge dt \tag{1.13}$$

$$\alpha_2 = dp \wedge dt - r\, dx \wedge dt \tag{1.14}$$

$$\alpha_3 = du \wedge dx - \frac{4}{3} d\omega \wedge dt \tag{1.15}$$

$$\alpha_4 = d\omega \wedge dx + \frac{3}{2} up\, dx \wedge dt + \frac{1}{4} dr \wedge dt \tag{1.16}$$

and can be shown [5] to have a prolongation defined by the two one forms

$$\Omega^1 = dg - \left(-\frac{3}{4} uf + \mu - \omega - gf\right)dx - \left(-\frac{1}{4} r - \frac{9}{16} u^2 + \frac{1}{4} pft(\omega-\mu)f + g^2\right)dt \tag{1.17}$$

$$\Omega^2 = df - \left(g - \frac{3}{4} u - f^2\right)dx - \left(\omega - \mu - \frac{1}{4} p + \frac{3}{4} uf + fg\right)dt \tag{1.18}$$

with two pseudopotentials g and f.

One can also have prolongations of ideals corresponding to equations in several spatial dimensions. For example, it can be shown [6] that the Kadomtsev-Petiashvili-Dryuma equation [7], [8]

$$\frac{3}{4}(u_{tt} + u_{xy}) + \frac{1}{4}(u_{xxx} + 6uu_x)_x = 0 \tag{1.19}$$

is equivalent to the ideal spanned by the four three-forms

$$\alpha_1 = du \wedge dt \wedge dy - p\, dx \wedge dt \wedge dy \tag{1.20}$$

$$\alpha_2 = dp \wedge dt \wedge dy - rdx \wedge dt \wedge dy \qquad (1.21)$$

$$\alpha_3 = du \wedge dx \wedge dy - \frac{4}{3} d\omega \wedge dt \wedge dy \qquad (1.22)$$

$$\alpha_4 = d\omega \wedge dx \wedge dy + \frac{3}{2} updx \wedge dt \wedge dy$$
$$+ \frac{1}{4} dx \wedge dt \wedge dy + \frac{3}{4} du \wedge dx \wedge dt \qquad (1.23)$$

and has a set of three prolongation two-forms defined by

$$\Omega^1 = \left(d\zeta^1 - \zeta^2 dx + \left(\zeta^3 + \frac{1}{4} u\zeta^1\right) dt \right) \wedge dy \qquad (1.24)$$

$$\Omega^2 = \left(d\zeta^2 - \left(\zeta^3 - \frac{3}{4} u\zeta^1\right) dx - \left[\frac{1}{2} u\zeta^2 + \left(\omega - \mu - \frac{1}{4} p\right)\zeta^1 \right] dt \right)$$
$$\wedge dy \wedge d\zeta^1 \wedge dt \qquad (1.25)$$

$$\Omega^3 = \left(d\zeta^3 + \left(\frac{3}{4} u\zeta^2 + (\omega-\mu)\zeta^1 \right) dx \right.$$
$$\left. + \left[\left(\frac{1}{4} r + \frac{9}{16} u^2 \right) \zeta^1 - \left(\omega - \mu + \frac{1}{4} p \right) \zeta^2 + \frac{1}{4} u\zeta^3 \right] dt \right) \wedge dy$$
$$+ 3dx \wedge d\zeta^1 - d\zeta^2 \wedge dt + 2\zeta^3 dx \wedge dt \qquad (1.26)$$

expressible in terms of three pseudopotentials ζ^1, ζ^2, and ζ^3.

2. BÄCKLUND TRANSFORMATIONS

If we denote the space to which the pseudopotentials belong by Y, then we define a <u>Bäcklund</u> transformation to be a diffeomorphism B

$$B: M \times Y \to M \times Y \qquad (2.1)$$

which has the property that

$B^*(I) \subset$ (Grassman ideal generated by I and Ω) .

This is a fairly strong definition that may require weakening. For example, we usually think of the transformation as associated with the equation rather than the ideal. There may be several different ideals giving rise, upon sectioning, to the same equation. For example, the ideal

$$\alpha_1' = du \wedge dt - pdx \wedge dt \qquad (2.2)$$

$$\alpha_2' = dp \wedge dt - rdx \wedge dt \qquad (2.3)$$

$$\alpha_3' = du \wedge dx + \frac{4}{3} d\omega \wedge dt \qquad (2.4)$$

$$\alpha_4' = -d\omega \wedge dx + \frac{3}{2} up \, dx \wedge dt + \frac{1}{4} dr \wedge dt \qquad (2.5)$$

obtained by replacing ω by $-\omega$ in (1.13)-(1.16) also gives rise to the Bousinesq equation (1.12). However, the definition above is adequate for the moment. Let us illustrate this definition by considering the Sine-Gordon equation. If we choose

$$A = \frac{i}{4\lambda} \cos u,$$

$$B = C = \frac{i}{4\lambda} \sin u, \qquad (2.7)$$

$$q = r = \frac{u_x}{2} = \frac{p}{2},$$

then the prlongation form Ω of Equation (1.10) is given by

$$\Omega = dy - \left[\frac{p}{2}(1+y^2) + 2i\lambda y\right]dx - \frac{i}{4\lambda}[\sin u(1-y^2) - 2\cos y]\,dt \qquad (2.8)$$

and the ideal I is spanned by the two forms

$$\alpha_1 = du \wedge dt - p\,dx \wedge dt \qquad (2.9)$$

$$\alpha_2 = dp \wedge dx + \sin u\, dx \wedge dt \qquad (2.10)$$

The form Ω can be converted to a simpler form by introducing the variable ϕ defined by

$$y = \tan \frac{\phi}{2}. \qquad (2.11)$$

A prolongation form is then given by ω defined by

$$\omega = d\phi - [p + 2i\lambda \sin \phi]dx - \frac{i}{2\lambda}[\sin(u-\phi)]dt \qquad (2.12)$$

BÄCKLUND TRANSFORMATIONS

We are seeking new values $u' = u'(u,p,\phi)$ and $p' = p'(u,p,\phi)$ which preserve the ideal spanned by α_1 and α_2 modulo elements of $F^*(M) \wedge \Omega$

$$\alpha_1' = du' \wedge dt - p'dx \wedge dt$$

$$= (\alpha_1 u_u' + u_\phi' \Omega \wedge dt) + dx \wedge dt[(u_u' + u_\phi')p + 2i\lambda u_\phi' \sin \phi - p']$$

$$+ (dp \wedge dt)u_p' \qquad (2.13)$$

$$\alpha_2' = dp' \wedge dx + \sin u' dx \wedge dt$$

$$= (\alpha_2 p_p' + p_\phi' \Omega \wedge dx) + dx \wedge dt\left[-p_p' \sin u - \frac{i}{2\lambda} \sin(u-\phi) + \sin u'\right]$$

$$+ (du \wedge dx)p_u' \qquad (2.14)$$

Consequently, we must have

$$u_p' = 0 \qquad (2.15)$$

$$p' = p(u_u' + u_\phi') + 2i\lambda u_\phi' \sin \phi \qquad (2.16)$$

$$p_u' = 0 \qquad (2.17)$$

$$p_p' \sin u + \frac{i}{2\lambda} \sin(u-\phi) p_\phi' - \sin u' \qquad (2.18)$$

Substituting (2.16) into (2.17) gives

$$u_{u\phi}' = 0 = u_{uu}' \quad , \qquad (2.19)$$

and so

$$u' = du + \beta\phi + \gamma \qquad (2.20)$$

$$p' = p(\alpha + \beta) + 2i\lambda\beta \sin \phi \qquad (2.21)$$

Putting these into (2.18) gives

$$(\alpha+\beta)\sin u + \beta \cos \phi \sin(\phi-u) = \sin(\alpha u+\beta\phi+\gamma) \qquad (2.22)$$

If we put $u = 0$, we see that

$$\beta \cos \phi \sin \phi = \sin(\beta\phi + \gamma) \qquad (2.23)$$

and so $\beta = 2$, $\gamma = 0$. Equation (2.22) becomes

$$(\alpha + 2)\sin u + 2 \cos \phi \sin(\phi - u) = \sin(\alpha u + 2\phi) \qquad (2.24)$$

which is satisfied by $\alpha = -1$. Therefore, the transformation $B_\lambda : M \times Y \to M \times Y$ defined by

$$(u,p,\phi) \to (-u + 2\phi, \ p + 4i\lambda \sin \phi, \phi) \qquad (2.25)$$

is a Bäcklund transformation for each value of λ. Similarly, it can be shown [9], [10] that

$$B_\lambda : M \times Y \to M \times Y$$

defined by

$$(\psi, P, y) \to \left(\psi - \frac{2i(\lambda-\bar{\lambda})\bar{y}}{(1+y\bar{y})} , \ P + 2i(\lambda-\bar{\lambda}) \frac{[\psi + 2i\bar{y}(\bar{\lambda}+\lambda y\bar{y})]}{(1+y\bar{y})^2} , \ y \right)$$

$$(2.26)$$

where $Y = \mathbb{C}$ is a Bäcklund transformation of the ideal I spanned by the forms

$$\alpha_1 = d\psi \wedge dt - P dx \wedge dt \qquad (2.27)$$

$$\alpha_2 = i d\psi \wedge dx + \frac{1}{2} dP \wedge dt + \psi|\psi|^2 dx \wedge dt \qquad (2.28)$$

and their complex conjugates and prolonged by the one-form

$$\Omega = dy + [\psi - y(2i\lambda - \psi y)] dx$$

$$+ \left[\frac{i}{2} P - \lambda\bar{\psi} + y\left(2i\lambda^2 - i|\psi|^2 - \frac{i}{2} Py - \lambda\psi y\right)\right] dt \qquad (2.29)$$

3. TOPOLOGICAL CHARGE [11]

The Sine-Gordon equation is invariant under the transformation

$$u \to u + 2\pi \qquad (3.1)$$

If $\chi = e^{iu}$, then $\chi: R \to S^1$ with the boundary constraints $\chi(\pm\infty) = 1$ is characterized by integers called windling numbers. In simple terms

$$u(+\infty) - u(-\infty) = 2\pi \times n \qquad (3.2)$$

The charge Q of a solution is defined to be the integer n. The conservation of this quantity is due entirely to its topological origin and is independent of the dynamics.

The equations which result from our Bäcklund transformation of Equation (2.25) give, upon sectioning onto a solution manifold of the prolonged ideal spanned by I and ω, the equation

$$u' = -u + 2\phi(u) \tag{3.3}$$

where $\phi(u)$ is a solution of the equations

$$\phi_x = p + 2i\lambda \sin \phi \tag{3.4}$$

$$\phi_t = \frac{i}{2\lambda} \sin(u - \phi) \tag{3.5}$$

When $u = 0$, (3.4) and (3.5) can be solved to give

$$\phi = 2 \tan^{-1}\left(\exp\left[2i\lambda x - \frac{i}{2\lambda} t\right]\right) \tag{3.6}$$

The soliton solutions correspond to imaginary eigenvalues $\lambda = ia/2$ and by (3.3) we obtain the soliton solution to the Sine-Gordon equation as

$$u^{sol}(x,t) = 4 \tan^{-1}[\exp(ax + a^{-1}t)] \tag{3.7}$$

For $a > 0$,

$$u^{sol}(\infty, t) - u^{sol}(-\infty, t) = 2\pi \quad ,$$

and so

$$Q^{sol} = 1 \quad . \tag{3.8}$$

It is interesting to note that the Bäcklund transformation

as expressed in (3.3)-(3.5) does not preserve topological charge.

$$Q_{u'} = -Q_u + 2Q_{\phi(u)} \qquad (3.9)$$

$$Q_{\phi(u)} = Q_{\phi(0)} = \frac{1}{2} Q^{sol} = \frac{1}{2} \qquad (3.10)$$

$$Q_{u'} = -Q_u + 1 \quad . \qquad (3.11)$$

A Bäcklund transformation has the effect of reversing the charge and adding an additional unit of charge. We think of a Bäcklund transformation as adding a soliton to a state, but, in fact, it does a little more, topologically speaking.

Another model which incorporates a topologically conserved quantity is the chiral pion Lagrangian [12] in one time and D space dimension, defined to be

$$L = \frac{1}{2} g_{\ddot{y}}(\pi) \partial_\mu \pi^i \partial^\mu \pi^i \quad , \qquad i = 1,2,3 \qquad (3.12)$$

where

$$g_{\ddot{y}}(\pi) = \delta_{ij} + \frac{\pi^i \pi^j}{(f^2 - \pi^2)} \qquad (3.13)$$

is a chirally invariant metric on the group space of $SU(2)$.

The problem with scalar field theories in more than one space dimension is that they run into the problem of Derrick's theorem [13] in one form or another. In this particular case of the chiral Lagrangian, the classical energy of the static field $\pi^i(x)$ is given by

$$E[\pi(x)] = \frac{1}{2} \int d^{D-1}x \ g_{ij}(\pi) \partial_\mu \pi^i \partial^\mu \pi^j \qquad (3.14)$$

and if we define $\pi^i_\lambda(x) = \pi^i(\lambda x)$ then from the equation of motion of the π^i field

$$\left(\Box + \frac{2}{f^2} L \right) \pi^k = 0 \qquad (3.15)$$

we see that $\pi^i_\lambda(x)$ is also a solution with energy

$$E[\pi_\lambda] = \lambda^{3-D} E[\pi] \qquad (3.16)$$

Consequently, given any static solution, we can always find another with lower <u>total</u> energy unless $D = 3$.

If a solution is sought for $D = 3$ in the form

$$\pi^1 = f \cos \beta(\phi) \sin \Theta(\rho) \qquad (3.17)$$

$$\pi^2 = f \sin \beta(\phi) \sin \Theta(\rho) \qquad (3.18)$$

$$\pi^3 = 0 \qquad (3.19)$$

where ρ and ϕ are polar coordinates in (x^1-x^2) space, then the equations of motion reduce to

$$\frac{d^2\psi}{dx^2} - n^2 \sin \psi = 0 \qquad (3.20)$$

where

$$\psi = 2\Theta, \qquad x = \ln \rho, \qquad \beta(\phi) = n\phi, \qquad (3.21)$$

BÄCKLUND TRANSFORMATIONS

and the soliton solutions of the Sine-Gordon equation give rise to the soliton solutions

$$\pi^1 = f \cos(n\phi) \frac{2\rho^n}{1 + \rho^{2n}} \tag{3.22}$$

$$\pi^2 = f \sin(n\phi) \frac{2\rho^n}{1 + \rho^{2n}} \tag{3.23}$$

$$\pi^3 = 0 \tag{3.24}$$

of the $D = 3$ chiral Lagrangian (3.12).

For fixed p these are mappings of physical space S' onto the isospin space S;

$$\pi_n : S' \to S'$$

The topological charge is given by

$$Q = \frac{1}{2\pi} \int_0^{2\pi} d\phi \, \varepsilon_{ij} \pi^i \frac{\partial}{\partial \phi} \pi^j$$

$$= n \quad , \qquad i = 1, 2 \tag{3.25}$$

and is usually called the <u>vortex number</u>.

4. THE MASSIVE THIRRING MODEL

A model which is of interest to workers in elementary particle theory is a two dimensional self interacting fermion field model described by the Lagrangian

$$L = \bar{\psi}(i\not{\partial} - m)\psi + \frac{1}{2} g J_\mu J^\mu \tag{4.1}$$

the current J_μ is given by

$$J_\mu = \bar{\psi}\gamma_\mu \psi \tag{4.2}$$

and the equations of motion which result are

$$i\not{\partial}\psi = m\psi - g\not{J}\psi \tag{4.3}$$

In the basis $\gamma^0 = \sigma_1$, $\gamma^1 = i\sigma_2$, the equations become

$$-i\psi_{1x} = \frac{1}{2}\psi_2 - 2\psi_2^+\psi_2\psi_1 \tag{4.4}$$

$$i\psi_{2t} = \frac{1}{2}\psi_1 - 2\psi_1^+\psi_1\psi_2 \tag{4.5}$$

A connection can be made between this equation and the Sine-Gordon equation which allows one to find soliton solutions for it.

If we section the prolongation form in Equation (2.8)

$$\Omega = dy - \left[\frac{p}{2}(1+y^2) + 2i\lambda y\right]dx$$

$$- \frac{i}{4\lambda}[\sin u(1-y^2) - 2\cos u\, y]dt \tag{4.6}$$

we obtain the equations

$$y_x = \frac{p}{2}(1+y^2) + 2i\lambda y \qquad (4.7)$$

$$y_t = \frac{i}{4\lambda}[\sin u(1-y^2) - 2\cos u\, y] \qquad (4.8)$$

These equations can be linearized if we change to projective coordinates χ_1 and χ_2 defined by

$$\frac{\chi_2}{\chi_1} = \frac{1+iy}{1-iy} \qquad (4.9)$$

or

$$y = \frac{i(\chi_1 - \chi_2)}{(\chi_1 + \chi_2)} \qquad (4.10)$$

Equations (4.7) and (4.8) then become

$$\chi_{1x} = \frac{a}{2}\chi_2 - \frac{ip}{2}\chi_1 \qquad (4.11)$$

$$\chi_{2x} = \frac{a}{2}\chi_1 + \frac{ip}{2}\chi_2 \qquad (4.12)$$

$$\chi_{1t} = \frac{1}{2a}e^{-iu}\chi_2 \qquad (4.13)$$

$$\chi_{2t} = \frac{1}{2a}e^{iu}\chi_1 \qquad (4.14)$$

where $\lambda = ia/2$.

If we define

$$\psi_1 = ia^{-1/2}\chi_1 \qquad (4.15)$$

$$\psi_2 = a^{1/2}\chi_2 \qquad (4.16)$$

(4.11) and (4.14) become

$$-i\psi_{1x} = \tfrac{1}{2}\psi_2 - \tfrac{1}{2}u_x\psi_1 \qquad (4.17)$$

$$i\psi_{2t} = \tfrac{1}{2}\psi_1 - \tfrac{1}{2}(1-e^{-iu})\psi_1 \qquad (4.18)$$

which are very like the Equations (4.4)-(4.5) of the massive Thirring model if we make the identifications

$$\psi_2^+\psi_2 \to \tfrac{1}{4}u_x \qquad (4.19)$$

$$\psi_1^+\psi_2 \to \tfrac{1}{4}(1-e^{iu}) \qquad (4.20)$$

To show that we can make such an identification, an alternative form of the Bäcklund transformation due to Hirota [16] may be used.

In Hirota's method we concentrate on solutions to the Sine-Gordon equation expressible in the form

$$u(x,t) = 4\tan^{-1}(f/g)$$
$$u' = 4\tan^{-1}(f'/g') \quad . \qquad (4.21)$$

BÄCKLUND TRANSFORMATIONS

The Bäcklund transformation can then be expressed in the form

$$(f' \pm ig')\overset{\leftrightarrow}{\partial}_x(f \pm ig) = \frac{u}{2}(f' \mp ig')(f \mp ig) \quad (4.22)$$

$$(f' \pm ig')\overset{\leftrightarrow}{\partial}_t(f \mp ig) = \frac{1}{2a}(f' \mp ig')(f \pm ig) \quad (4.23)$$

where

$$A\overset{\leftrightarrow}{\partial}_x B = AB_x - BA_x \quad .$$

From these one can show that χ_1 and χ_2 defined by

$$\chi_1 = \frac{f - ig}{f' + ig'} \quad (4.24)$$

$$\chi_2 = \frac{f + ig}{f' - ig'} \quad (4.25)$$

are solutions of Equations (4.11)-(4.14).

To find solutions to the massive Thirring model we will have to "invert the Bäcklund transformation". The one solution is

$$u'_a = 4 \tan^{-1}(\exp[ax + a^{-1}t])$$
$$= 4 \tan^{-1}(g_a'^{sol}/f_a'^{sol}) \quad (4.26)$$

The initial solution is

$$u = 0 = 4 \tan^{-1}(g^0/f^0) \quad . \quad (4.27)$$

If we choose $f^0 = 1$, $g^0 = 0$, then, $\chi_1 = (f_a'^{sol} + ig_a'^{sol})^{-1}$
$\chi_2 = (f_a'^{sol} - ig_a'^{sol})^{-1}$. From these equations we can show that

$$a\chi_2^+\chi_2 = \frac{1}{4} u_{ax}'^{sol} \tag{4.28}$$

$$a^{-1}\chi_1^+\chi_1 = \frac{1}{4} u_{at}'^{sol} \tag{4.29}$$

$$-i\chi_1^+\chi_2 = \frac{1}{4}\left(1 - e^{iu_a'^{sol}}\right) \tag{4.30}$$

which include the identifications required in (4.19)-(4.20). Therefore, an explicit solution to the Thirring model which exactly implements the Coleman correspondences [14].

$$:\psi_1^+\psi_2: \to \frac{1}{4} u_x \tag{4.31}$$

$$:\psi_1^+\psi_1: \to \frac{1}{4} u_t \tag{4.32}$$

$$:\psi_1^+\psi_2: \to \frac{1}{4}(1 - e^{iu}) \tag{4.33}$$

is given by

$$\psi_1 = ia^{-1/2}\left(\frac{1}{2}\sin u_a'^{sol}/2\right)^{1/2} e^{-iu_a'^{sol}/2} \tag{4.34}$$

$$\psi_2 = a^{1/2}\left(\frac{1}{2}\sin u_a'^{sol}/2\right)^{1/2} e^{iu_a'^{sol}/2} \tag{4.35}$$

BIBLIOGRAPHY

1. W. Slebodzinski, Exterior forms and applications (Polish Scientific Publishers, Warsaw, 1970).

2. M.J. Ablowitz, D.J. Kaup, A.C. Newell, and H. Segur, Phys. Rev. Lett. **31**, 125 (1973).

3. H.D. Wahlquist and F.B. Estabrook, J. Math. Phys. **16**, 1, (1975).

4. H.C. Morris, A prolongation structure for the AKNS system and its generalisation, Trinity College Preprint TCD-1976-6, J. Math. Phys. (in press).

5. H.C. Morris, Prolongation structures and a generalised inverse scattering problem, Trinity College Preprint TCD 1975-6, J. Math. Phys. (in press).

6. H.C. Morris, Prolongation structures and nonlinear evolution equations in two spatial dimensiona, Trinity College Preprint TCD 1975-10, J. Math. Phys. (in press).

7. B. Kodomtsev and V. Petviashvili, Soviet Physics Dokl. **15**, 539 (1970).

8. V. Dryuma, J.E.T.P. Lett. **19**, 387 (1974).

9. H.D. Wahlquist and F.B. Estabrook, Prolongation structures of nonlinear evolution equations II, The nonlinear Schrodinger equation, Caltech preprint, 1975.

10. J. Corones, J. Math. Phys. **17**, 756 (1976).

11. A. Patani, M. Schlindwein and Q. Shafi, Topological charges in field theory, Univ. of Freiburg preprint (1976).

12. M.J. Duff and C.J. Isham, Soliton on vortex type solutions in non linear chiral theories, Preprint Kings College, London 1976.

13. G.H. Derrick, J. Math. Phys. **5**, 1252 (1964).

14. S. Coleman, Phys. Rev. D**11**, 2088 (1975).

15. S. Orfanidis, Soliton solutions of the massive Thirring model and the inverse scattering transform, New York Univ. preprint NYU/TR1/76 (1976).

16. R. Hirota, Progr. Theor. Phys. 52, 1498 (1975).

PROLONGATION STRUCTURES OF NONLINEAR EVOLUTION
EQUATIONS IN TWO AND THREE SPATIAL DIMENSIONS

Hedley C. Morris
Trinity College
Dublin, Ireland

INTRODUCTION

In 1975 Frank Estabrook and Hugo Wahlquist [1] introduced a new geometric approach to the study of two dimensional systems soluble by the inverse scattering technique. Their method was to use the results of E. Cartan and express partial differential equations such as that of Korteweg and de Vries by an equivalent closed ideal of differential forms. This was an extension of earlier work by Estabrook and Harrison [2] and details may be found in Professor Estabrook's lectures in this volume.

In this article I would like to discuss a possible way in which the prolongation structure technique might be extended to evolution equations in three or more dimensions. It is by no means the most general that is possible within the geometric framework of Wahlquist and Estabrook, but is simply the most obvious. In view of its simplicity it is encouraging to find that it can yield the Kadomtsev-Petviashvili-Dryuma equation [3], [4], and wave-envelope equations [5] as well as other equations of as yet unknown physical significance.

Hopefully, further development in the near future will allow a geometric treatment of the Benny equations [6] which are a prime candidate for soliton behavior in view of the existence of the infinite set of local conserved quantities determined by Miura [7]. The massive Thirring model with anticommuting fields [8] is also known to possess an infinite set of conserved quantities and I would expect a generalized version of the prolongation method to yeild results for these equations also.

In Section 1 an extension of the prolongation structure approach to more than two variables is outlined in order to introduce notation. In the following three sections, we will illustrate the method by applying it in numerous cases. Finally, in Section 5 we indicate how the method can be further applied to produce equations in three spatial variables which have prolongation structures.

1. GENERALIZING A KNOWN CASE

The differential forms which are used to represent a given differential equation are of two basic types. There are forms which are introduced in order to reduce the

PROLONGATION STRUCTURES

equation to a set of first order equations and those which represent the equation itself. The former type we call linearizing forms and the latter we call dynamic forms.

In the Case of the K.dV equation the ideal is taken to be [1]

$$\alpha_1 = du \wedge dt - pdx \wedge dt \qquad (1.1)$$

$$\alpha_2 = dp \wedge dt - rdx \wedge dt \qquad (1.2)$$

$$\alpha_3 = dr \wedge dt - du \wedge dx + 12up\, dx \wedge dt \qquad (1.3)$$

and on sectioning these forms onto the solution submanifold, we find that

$$\tilde{\alpha}_1 = 0 \quad \text{gives} \quad p = u_x \qquad (1.4)$$

$$\tilde{\alpha}_3 = 0 \quad \text{gives} \quad r = p_x = u_{xx} \qquad (1.5)$$

and

$$\tilde{\alpha}_3 = 0 \quad \text{gives the K.dV equation}$$

$$u_t + u_{xxx} + 12uu_x = 0 \qquad (1.6)$$

Thus, for the K.dV equation expressed as the ideal (1.1)-(1.3), α_1 and α_2 are the linearizing forms and α_3 is the dynamic form.

Let us introduce some notation for future use. Suppose that $\{\alpha_i\}_1^N$ are a closed set of two-forms. This means that there are N^2 forms K_i^e having the property that

$$d\alpha_i = \sum_{e=1}^{N} K_i^e \wedge \alpha_e \quad , \quad i = 1,\ldots,N \qquad (1.7)$$

Suppose also that this ideal has a linear prolongation structure

$$\Omega^\beta = \sum_{\alpha=1}^{M} (F_\alpha^\beta dx + G_\alpha^\beta dt)\zeta^\alpha + d\zeta^\beta \qquad (1.8)$$

where $\underline{F}$ and $\underline{G}$ are matrices independent of $\underline{\zeta}$. This means that we can find functions $f^{\beta i}$ and one-forms η_γ^β such that

$$d\Omega^\beta = \sum_{i=1}^{N} f^{\beta i} \alpha_i + \sum_{\gamma=1}^{M} \eta_\gamma^\beta \wedge \Omega^\gamma \qquad (1.9)$$

Let us also suppose that the $\{\alpha_i\}$ are labelled so that

$\alpha_i, \quad i = 1,\ldots,K \quad$ are the <u>linearizing forms</u>

and

$\alpha_j, \quad j = K+1,\ldots,N \quad$ are the <u>dynamic forms</u>

It is possible to generalize the ideal $\{\alpha_i\}_1^N$ to forms in more than two spatial dimensions in such a way that the new ideal still has a prolongation structure. In the following section we consider one possible way in which such a result can be obtained.

2. AN EXTENDED IDEAL

Let us seek a new ideal consisting of three-forms which is closed and represents an extension of the ideal spanned by α_i ($i = 1,\ldots,N$) to higher dimensions. The extension is to be constructed in such a way that the new ideal also possesses a prolongation form $\bar{\Omega}$. The ideal spanned by the three-forms

$$\bar{\alpha}_i = \alpha_i \wedge dy, \qquad i = 1,\ldots,K \qquad (2.1)$$

$$\bar{\alpha}_j = \alpha_j \wedge dy + \beta_j, \qquad j = K+1,\ldots,N \qquad (2.2)$$

will be closed so long as the β_j have the property

$$d\beta_j = \sum_{k=K+1}^{N} K_j^k \wedge \beta_k \qquad (2.3)$$

We attempt to construct a prolongation form $\bar{\Omega}$ which will produce such an ideal. As an ansatz we seek $\bar{\Omega}$ in the form

$$\bar{\Omega} = \Omega \wedge dy + \underset{\sim}{H} dx \wedge dt + (\underset{\sim}{A}dx + \underset{\sim}{B}dt) \wedge d\zeta \qquad (2.4)$$

or

$$\bar{\Omega} = d\zeta \wedge dy + F\zeta dx \wedge dy + G\zeta dt \wedge dy + H\zeta dx \wedge dt + Adx \wedge d\zeta + Bdt \wedge d\zeta$$

$$= (Fdx \wedge dy + Gdt \wedge dy + Hdx \wedge dt)\zeta + (Adx + Bdt - dy) \wedge d\zeta, \qquad (2.5)$$

where A and B are constant $M \times M$ matrices. It can be shown that

$$d\bar{\Omega}^\beta = \sum_{k=1}^{N} f^{\beta_i} \bar{\alpha}_i + \sum_{\gamma=1}^{M} \eta_\gamma^\beta \wedge \bar{\Omega}^\gamma \qquad (2.6)$$

provided that the matrix H is given by

$$\underset{\sim}{H} = \underset{\sim\sim}{GA} - \underset{\sim\sim}{FB} \qquad (2.7)$$

and

$$\sum_{i=K+1}^{N} f^{\beta_i} \beta_i = (dGA - dFB)^\beta \wedge dx \wedge dt \qquad (2.8)$$

Such β_i will trivially satisfy the condition (2.3). If we section $\bar{\Omega}^\beta$ onto a solution manifold of the extended ideal, we obtain

$$\zeta_x = -F\zeta - A\zeta_y \qquad (2.9)$$

$$\zeta_t = -G\zeta - G\zeta_y \qquad (2.10)$$

$$A\zeta_t - B\zeta_x = -H\zeta \qquad (2.11)$$

which require for consistency that

$$[A,B] = 0 \qquad (2.12)$$

$$[G,A] + [B,F] = 0 \qquad (2.13)$$

PROLONGATION STRUCTURES

Equations (2.4), (2.7)-(2.8) and (2.12)-(2.13) are the basic equations of our approach.

We can link our method to the results of Zakharov and Shabat [9] if we note that the operators

$$L^1 = A \frac{\partial}{\partial y} + F$$
$$L^2 = B \frac{\partial}{\partial y} + G \qquad (2.14)$$

give rise to the generalized Lax representation

$$L^1_t - L^2_x = [L^1, L^2] \qquad (2.15)$$

for the extended equations. However, with the exception of the wave-envelope equations, this form of representation is not particularly interesting for the degenerate types of A and B involved here. In the following section we illustrate this technique by showing how the Kadomtsev-Petviashvili-Dryuma equation results from its application to the Boussinesq equation.

3. THE KADOMTSEV-PETVIASHVILI-DRYUMA EQUATION

A prolongation structure for the Boussinesq equation

$$\frac{3}{4} u_{tt} + \frac{1}{4} u_{xxxx} + \frac{3}{2} (uu_x)_x = 0 \qquad (3.1)$$

expressed in terms of the four two-forms

$$\alpha_1 = du \wedge dt - pdx \wedge dt \qquad (3.2)$$

$$\alpha_2 = dp \wedge dt - rdx \wedge dt \qquad (3.3)$$

$$\alpha_3 = du \wedge dx - \frac{4}{3} dw \wedge dt \qquad (3.4)$$

$$\alpha_4 = dw \wedge dx + \frac{3}{2} up\, dx \wedge dt + \frac{1}{4} dr \wedge dt \qquad (3.5)$$

is given by [10]

$$\Omega^1 = d\zeta^1 - \zeta^2 dx + \left(\zeta^3 + \frac{1}{4} u\zeta^1\right) dt \qquad (3.6)$$

$$\Omega^2 = d\zeta^2 - \left(\zeta^3 - \frac{3}{4} u\zeta^1\right) dx - \left(\frac{1}{2} u\zeta^2 + \left(w - \mu - \frac{1}{4} p\right)\zeta^1\right) \qquad (3.7)$$

$$\Omega^3 = d\zeta^3 + \left(\left(\frac{3}{4} u\zeta^2 + (w - \mu)\zeta^1\right) dx \right.$$
$$\left. + \left(\frac{1}{4} r + \frac{9}{16} u\right)\zeta^1 - \left(w - \mu + \frac{1}{4} p\right)\zeta^2 + \frac{1}{4} u\zeta^3\right) dt \qquad (3.8)$$

α_1, α_2 and α_3 are <u>linearization forms</u> and α_4 is a <u>dynamic form</u>. Equation (**) takes the form

$$\overbrace{\begin{bmatrix} 0 & 0 & 0 \\ 0 & 0 & 0 \\ 1 & 0 & 0 \end{bmatrix}}^{f_4} \beta_4$$

$$= \begin{bmatrix} \frac{1}{4} dy \wedge dx \wedge dt & \overbrace{0}^{dG} & 0 \\ \left(\frac{1}{4} dp - dw\right) \wedge dx \wedge dt, & 0 & 0 \\ \left(\frac{1}{4} dr + \frac{9}{8} udu\right) \wedge dx \wedge dt, & \frac{3}{4} dt \wedge dx \wedge dt, & \frac{1}{4} du \wedge dx \wedge dt \end{bmatrix} A$$

(3.9)

$$- \begin{bmatrix} 0 & 0 & 0 \\ \frac{3}{4} du \wedge dx \wedge dt, & 0 & 0 \\ dw \wedge dx \wedge dt, & \underbrace{\frac{3}{4} du \wedge dx \wedge dt}_{dF} & 0 \end{bmatrix} B$$

This equation has essentially only one solution:

$$\beta_4 = \frac{3}{4} du \wedge dx \wedge dt \tag{3.10}$$

and a pair of matrices which yield this result is given by

$$A = \frac{3}{4} \begin{bmatrix} 0 & 0 & 0 \\ 0 & 0 & 0 \\ 1 & 0 & 0 \end{bmatrix},$$

$$B = \left(-\frac{3}{4}\right) \begin{bmatrix} 0 & 0 & 0 \\ 1 & 0 & 0 \\ 0 & 1 & 0 \end{bmatrix}$$

The new set of forms

$$\bar{\alpha}_1, \bar{\alpha}_2, \bar{\alpha}_3, \bar{\alpha}_4 = \alpha_4 \wedge dy + \frac{3}{4} du \wedge dx \wedge dt \qquad (3.11)$$

correspond to the <u>Kadomtsev-Petviashvili-Dryuma</u> (3.4) equation.

$$\frac{3}{4}(u_{tt} + u_{xy}) + \frac{1}{4}(u_{xxx} + 6uu_x)_x = 0 \qquad (3.12)$$

We can now section the new prolongation forms $\bar{\Omega}$ onto a solution manifold of this equation and obtain the inverse scattering equations

$$\zeta^1_x = \zeta^2 \qquad (3.13)$$

$$\zeta^2_x = \zeta^3 - \frac{3}{4} u \zeta^1 \qquad (3.14)$$

$$\zeta^3_x = -\frac{3}{4} u \zeta^2 - (w - \mu)\zeta^1 - \frac{3}{4} \zeta^1_y \qquad (3.15)$$

$$\zeta^1_t = -\zeta^3 - \frac{1}{4} u \zeta^1 \qquad (3.16)$$

$$\zeta^2_t = \frac{1}{2} u \zeta^2 + \left(w - \mu - \frac{1}{4} p\right)\zeta^1 + \frac{3}{4} \zeta^1_y \qquad (3.17)$$

$$\zeta^3_t = \left(-\frac{1}{4} r - \frac{9}{16} u^2\right)\zeta^1 + \left(w - \mu + \frac{1}{4} p\right)\zeta^1$$
$$\qquad (3.18)$$
$$- \frac{1}{4} u \zeta^3 + \frac{3}{4} \zeta^3_y$$

Eliminating ζ^2 and ζ^3 between these equations, we can obtain an inverse scattering problem in ζ^1 alone

$$\frac{3}{4} \zeta^1_y + \zeta^1_{xxx} + \frac{3}{2} u\zeta^1_x + \left(\frac{3}{4} u_x + w\right)\zeta^1 = \mu\zeta^1 \qquad (3.19)$$

$$\zeta^1_t + \zeta^1_{xx} + u\zeta^1 = 0 \qquad (3.20)$$

This scattering problem was previously obtained by different methods by Zakharov and Shabat [10].

4. A GENERALIZED NONLINEAR SCHRÖDINGER EQUATION

The set of equations

$$\left(\frac{\partial^2}{\partial x^2} + \frac{\partial^2}{\partial y^2}\right) A = 2A(\Phi - \Psi) \qquad (4.1)$$

$$\left(\frac{\partial}{\partial y} - \frac{\partial}{\partial x}\right) \Phi = -\frac{1}{2} \left(\frac{\partial}{\partial x} + \frac{\partial}{\partial y}\right) (AA^*) \qquad (4.2)$$

$$\left(\frac{\partial}{\partial y} + \frac{\partial}{\partial x}\right) \Psi = \frac{1}{2} \left(\frac{\partial}{\partial y} - \frac{\partial}{\partial x}\right) (AA^*) \qquad (4.3)$$

expressed by the eight two-forms

$$\alpha_1 = dA \wedge dy - (R - L)dx \wedge dy ,$$
$$\alpha^*_1 = dA^* \wedge dy - (R^* - L^*)dx \wedge dy \qquad (4.4)$$

$$\alpha_2 = dA \wedge dx + (R+L)dx \wedge dy \quad , \tag{4.5}$$

$$\alpha_2^* = dA^* \wedge dx + (R^*+L^*)dx \wedge dy$$

$$\alpha_3 = d\Phi \wedge (dx+dy) - (RA^* + AR^*)dx \wedge dy \tag{4.6}$$

$$\alpha_4 = d\Psi \wedge (dx-dy) + (LA^* + AL^*)dx \wedge dy \tag{4.7}$$

$$\alpha_5 = dL \wedge dx - dR \wedge dy + A(\Phi-\Psi)dx \wedge dy \tag{4.8}$$

$$\alpha_6 = dR^* \wedge dx - dL^* \wedge dy + A^*(\Phi-\Psi)dx \wedge dy \tag{4.9}$$

has a prolongation structure

$$\Omega^1 = d\zeta^1 - (\lambda\zeta^1 + A\zeta^2 + \zeta^3)dx - \zeta^3 dy \tag{4.10}$$

$$\Omega^2 = d\zeta^2 - (A^*\zeta^1 + \lambda\zeta^2 - \zeta^4)dx - \zeta^4 dy \tag{4.11}$$

$$\Omega^3 = d\zeta^3 - (\Phi\zeta^1 + L\zeta^2 + \lambda\zeta^3)dx - (\Phi\zeta^1 - R\zeta^2 - A\zeta^4)dy \tag{4.12}$$

$$\Omega^4 = d\zeta^4 - (R^*\zeta^1 + \Psi\zeta^2 + \lambda\zeta^4)dx - (L^*\zeta^1 - \Psi\zeta^2 + A^*\zeta^3)dy \tag{4.13}$$

which gives rise to the scattering problem

$$\zeta_x = \begin{bmatrix} \lambda & A & 1 & 0 \\ A^* & \lambda & 0 & -1 \\ \Phi & L & \lambda & 0 \\ R^* & \Phi & 0 & \lambda \end{bmatrix} \zeta \qquad (4.14)$$

$$\zeta_y = \begin{bmatrix} 0 & 0 & 1 & 0 \\ 0 & 0 & 0 & 1 \\ \Phi & -R & 0 & -A \\ L^* & -\psi & A^* & 0 \end{bmatrix} \zeta \qquad (4.15)$$

The <u>dynamic forms</u> are α_3, α_4, α_5 and α_6 and Equation (2.8) becomes

$$-\begin{bmatrix} 0 & 0 & 0 & 0 \\ 0 & 0 & 0 & 0 \\ \beta_3 & \beta_5 & 0 & 0 \\ \beta_6 & \beta_4 & 0 & 0 \end{bmatrix} = (dGA - dFB) \wedge dx \wedge dy \qquad (4.16)$$

The matrices

$$A = \frac{i}{2} \begin{bmatrix} 0 & 0 & 0 & 0 \\ 0 & 0 & 0 & 0 \\ 1 & 0 & 0 & 0 \\ 0 & 1 & 0 & 0 \end{bmatrix}, \qquad (4.17a)$$

$$B = \frac{i}{2} \begin{bmatrix} 0 & 0 & 0 & 0 \\ 0 & 0 & 0 & 0 \\ 1 & 0 & 0 & 0 \\ 0 & -1 & 0 & 0 \end{bmatrix} \quad (4.17b)$$

satisfy the required conditions

$$[A,B] = 0 \quad (4.18)$$

$$[G,A] + [B,F] = 0$$

and the resulting forms β_2, β_4, β_5 and β_6 are given by

$$\beta_3 = 0$$

$$\beta_4 = 0 \quad (4.20)$$

$$\beta_5 = -\frac{i}{2} dA \wedge dx \wedge dt$$

$$\beta_6 = \frac{i}{2} dA^* \wedge dx \wedge dy$$

corresponding to the generalized <u>dynamic forms</u>

$$\bar{\alpha}_5 = \alpha_5 \wedge dt - \frac{i}{2} dA \wedge dx \wedge dy \quad (4.21)$$

$$\bar{\alpha}_6 = \alpha_6 \wedge dt + \frac{i}{2} dA^* \wedge dx \wedge dt \quad (4.22)$$

The new ideal spanned by $\bar{\alpha}_1$, $\bar{\alpha}_2$, $\bar{\alpha}_1^*$, $\bar{\alpha}_2^*$, $\bar{\alpha}_3$, $\bar{\alpha}_4$, $\bar{\alpha}_5$, $\bar{\alpha}_6$ is equivalent to the <u>generalized nonlinear Schrödinger equation</u>

PROLONGATION STRUCTURES

$$-i \frac{\partial}{\partial t} A = \nabla^2 A - 2A(\Phi - \Psi) \tag{4.23}$$

$$\left(\frac{\partial}{\partial y} - \frac{\partial}{\partial x}\right) \Phi = -\frac{1}{2}\left(\frac{\partial}{\partial x} + \frac{\partial}{\partial y}\right)(AA^*) \tag{4.24}$$

$$\left(\frac{\partial}{\partial y} + \frac{\partial}{\partial x}\right) \psi = \frac{1}{2}\left(\frac{\partial}{\partial y} - \frac{\partial}{\partial x}\right)(AA^*) \tag{4.25}$$

We can introduce a further parameter μ by replacing F and G by F + A and G + A which, as a result of Equations (2.12) and (2.13), will produce an equivalent prolongation structure. The generalized prolongation two-form Ω may then be sectioned onto a solution manifold of these equations to yield the inverse scattering problem

$$\zeta_x^1 = \lambda \zeta^1 + A\zeta^2 + \zeta^3 \tag{4.26}$$

$$\zeta_y^1 = \zeta^3 \tag{4.27}$$

$$\zeta_x^2 = A^*\zeta^1 + \lambda \zeta^2 - \zeta^4 \tag{4.28}$$

$$\zeta_y^2 = \zeta^4 \tag{4.29}$$

$$\zeta_x^3 = (\mu + \Phi)\zeta^1 + L\zeta^2 + \lambda \zeta^3 - \frac{i}{2}\zeta_t^1 \tag{4.30}$$

$$\zeta_y^3 = (\mu + \Phi)\zeta^1 - R\zeta^2 - A\zeta^4 - \frac{i}{2}\zeta_t^1 \tag{4.31}$$

$$\zeta_x^4 = R^*\zeta^1 + (\Psi + \mu)\zeta^2 + \lambda\zeta^4 - \frac{i}{2}\zeta_t^2 \qquad (4.32)$$

$$\zeta_y^4 = L^*\zeta^1 - (\Psi + \mu)\zeta^2 + A^*\zeta^3 + \frac{i}{2}\zeta_t^2 \qquad (4.33)$$

This inverse scattering problem has also been considered by <u>Ablowitz and Haberman</u> using different methods [13].

A slightly more compact form is obtained if we eliminate ζ^3 and ζ^4, and is given by

$$\zeta_x^1 = \lambda\zeta^1 + A\zeta^2 + \zeta_y^1 \qquad (4.34)$$

$$\zeta_x^2 = A^*\zeta^1 + \lambda\zeta^2 - \zeta_y^2 \qquad (4.35)$$

$$\frac{i}{2}\zeta_t^1 = (\mu + \Phi)\zeta^1 - R\zeta^2 - A\zeta_y^2 - \zeta_{yy}^1 \qquad (4.36)$$

$$\frac{i}{2}\zeta_t^2 = -L^*\zeta^1 + (\mu + \Psi)\zeta^2 - A^*\zeta_y^1 + \zeta_{yy}^2 \qquad (4.37)$$

If $\zeta_x^1 = 0 = \zeta_x^2$, then these equations reduce to

$$\zeta_y^1 = -\lambda\zeta^1 - A\zeta^2 \qquad (4.38)$$

$$\zeta_y^2 = A^*\zeta^1 + \lambda\zeta^2 \qquad (4.39)$$

$$\frac{i}{2}\zeta_t^1 = -\left(\frac{1}{2}AA^* + \lambda^2\right)\zeta^1 + \left(\frac{1}{2}A_y - \lambda A\right)\zeta^2 \qquad (4.40)$$

$$\frac{i}{2}\zeta_t^2 = \left(\frac{1}{2}A_y^* + \lambda A^*\right)\zeta^1 + \left(\frac{1}{2}AA^* + \lambda^2\right)\zeta^2 \qquad (4.41)$$

PROLONGATION STRUCTURES 141

which are equivalent to the normal <u>Zakharov-Shabat</u> [14] equations for the nonlinear Schrödinger equation

$$i \frac{\partial}{\partial t} A + \frac{\partial^2}{\partial y^2} A + 2A|A|^2 = 0 \qquad (4.42)$$

In the Equations (4.34)-(4.37) it is clear that μ is the more useful parameter and so we put $\lambda = 0$.

Equations (4.34)-(4.37) can then be expressed in the matrix form

$$\left[I \frac{\partial}{\partial x} - \sigma \frac{\partial}{\partial y} - \begin{pmatrix} A & 0 \\ A^* & 0 \end{pmatrix} \right] \begin{pmatrix} \zeta^1 \\ \zeta^2 \end{pmatrix} = 0 \qquad (4.43)$$

and

$$\left[\frac{i}{2} I \frac{\partial}{\partial t} + \sigma \frac{\partial^2}{\partial y^2} + \begin{pmatrix} 0 & A \\ A^* & 0 \end{pmatrix} \frac{\partial}{\partial y} - \begin{pmatrix} \Phi & -R \\ -L^* & \Psi \end{pmatrix} \right] \begin{pmatrix} \zeta^1 \\ \zeta^2 \end{pmatrix} = \mu \begin{pmatrix} \zeta^1 \\ \zeta^2 \end{pmatrix}$$

(4.44)

where

$$I = \begin{bmatrix} 1 & 0 \\ 0 & 1 \end{bmatrix} \quad \text{and} \quad \sigma = \begin{bmatrix} 1 & 0 \\ 0 & -1 \end{bmatrix} \qquad (4.45)$$

The general inverse scattering problem of this form is

$$\left[I \frac{\partial}{\partial x} - \sigma \frac{\partial}{\partial y} - \begin{pmatrix} 0 & A \\ B & 0 \end{pmatrix} \right] \begin{pmatrix} \zeta^1 \\ \zeta^2 \end{pmatrix} = 0 \qquad (4.46)$$

and

$$\left[\frac{i}{2} I \frac{\partial}{\partial t} + \sigma \frac{\partial^2}{\partial y^2} + \begin{pmatrix} 0 & A \\ B & 0 \end{pmatrix} \frac{\partial}{\partial y} - \begin{pmatrix} \Phi & -R \\ -S & \Psi \end{pmatrix}\right] \begin{pmatrix} \zeta^1 \\ \zeta^2 \end{pmatrix} = \mu \begin{pmatrix} \zeta^1 \\ \zeta^2 \end{pmatrix}$$

(4.47)

where

$$S = \frac{1}{2} (\partial_y - \partial_x) B \qquad (4.48)$$

which is the inverse scattering problem appropriate to the equations

$$-\frac{i}{2} \frac{\partial}{\partial t} A = \nabla^2 A - 2A(\Phi - \Psi) \qquad (4.49)$$

$$-\frac{i}{2} \frac{\partial}{\partial t} B = \nabla^2 B - 2B(\Phi - \Psi) \qquad (4.50)$$

$$\left(\frac{\partial}{\partial y} - \frac{\partial}{\partial x}\right) \Phi = -\frac{1}{2} \left(\frac{\partial}{\partial x} + \frac{\partial}{\partial y}\right) (AB) \qquad (4.51)$$

$$\left(\frac{\partial}{\partial y} + \frac{\partial}{\partial x}\right) \Psi = \frac{1}{2} \left(\frac{\partial}{\partial y} - \frac{\partial}{\partial x}\right) (AB) \qquad (4.52)$$

Solutions of these equations, for which $\zeta = A^*$, provide solutions of our generalized Schrodinger equation. Solving the Gelfand-Levitan equations appropriate to (4.49)-(4.52) gives the single soliton solution

$$A = e^{4i(k-K)^2 t - 2(k-K)x} \, 2Ke^{4iK^2 t} \, \text{sech}(2Ky + \delta_0) \quad (4.53)$$

$$B = e^{-4i(K-K)^2 t + 2(k-K)x} \, 2Ke^{-4iK^2 t} \, \text{sech}(2Ky + \delta_0)$$

(4.54)

$$\Phi = -\frac{1}{2}(AB)$$

$$= -\Psi \qquad (4.55)$$

$$= 4K^2 \operatorname{sech}^2(2Ky + \delta_0)$$

Only in the case $k = K$ is $B = A^*$ and this is the normal soliton of the nonlinear Schrodinger equation

$$i\frac{\partial}{\partial t}A + \frac{\partial^2}{\partial y^2}A + 2|A|^2 A = 0 \qquad (4.56)$$

Thus, for this case, unlike the K-P-D equation, there would not appear to be generalized single soliton solution.

5. THREE DIMENSIONS

We can easily extend the method we have developed to include additional spatial dimensions.

If we take the prolongation form $\overline{\Omega}$ corresponding to the ideal spanned by $\{\overline{\alpha}_i\}_{i=1}^N$, then

$$\overline{\overline{\Omega}} = \overline{\Omega} \wedge dz + I\zeta dx \wedge dy \wedge dt + (Cdx \wedge dt + Ddx \wedge dy + Edt \wedge dy) \wedge d\zeta \qquad (5.1)$$

is a prolongation structure for the ideal spanned by

$$\overline{\overline{\alpha}}_i = \overline{\alpha}_i \wedge dz, \qquad i = 1, \ldots, K \qquad (5.2)$$

$$\overline{\overline{\alpha}}_j = \overline{\alpha}_j \wedge dz + \gamma_j, \qquad j = K+1, \ldots, N \qquad (5.3)$$

if

$$[D,E] = 0 \tag{5.4}$$

$$[G,D] + [E,F] = 0 \tag{5.5}$$

and the matrices C and I are given by

$$C = EA - DB \tag{5.6}$$

$$I = GD - FE \tag{5.7}$$

If these equations are satisfied, the extending forms γ_i are determined by

$$\sum_{i=K+1}^{N} f^{\beta_i} \gamma_i = (dGD - dFE) \wedge dx \wedge dy \wedge dt \tag{5.8}$$

and the extended ideal $\{\bar{\bar{\alpha}}_i\}_{i=1}^{N}$ has a prolongation structure

$$\begin{aligned}\bar{\bar{\Omega}} = {} & \bar{\Omega} \wedge dz + (GD - FE)\zeta dx \wedge dy \wedge dt + (EA - DB)dx \wedge dt \wedge d\zeta \\ & + (Ddx \wedge dy + Edt \wedge dy) \wedge d\zeta\end{aligned} \tag{5.9}$$

giving rise, upon sectioning, to the inverse scattering problem

$$\zeta_x = -F\zeta - A\zeta_y - D\zeta_z \tag{5.10}$$

$$\zeta_t = -G\zeta - B\zeta_y - E\zeta_z \tag{5.11}$$

PROLONGATION STRUCTURES

An example of such a three dimensional system is provided by the nonlinear wave envelope equations. These equations can be expressed in the matrix form

$$N_t = (\alpha N)_x + [\alpha N, N] \tag{5.12}$$

where N is an $(n \times n)$ matrix and

$$(\alpha N)_{ij} \stackrel{def}{=} \alpha_{ij} N_{ij} .$$

It has been shown that by using the method we have developed an inverse scattering problem can be developed for the equation

$$N_t = (\alpha N)_x + (\gamma N)_y + [\alpha N, N] \tag{5.13}$$

Clearly, we have just shown that

$$N_t = (\alpha N)_x + (\gamma N)_y + (\delta N)_z + [\alpha N, N] \tag{5.14}$$

also has such an inverse scattering problem. Also, the equation

$$\frac{3}{4}(u_{tt} + u_{xy} + u_{xz}) + \frac{1}{4}(u_{xxx} + 6uu_x)_x = 0 \tag{5.15}$$

clearly has the inverse scattering formulation

$$\frac{3}{4}(\zeta_y^1 + \zeta_z^1) + \zeta_{xxx}^1 + \frac{3}{2} u \zeta_x^1 + \left(\frac{3}{4} u_x + w\right)\zeta^1 = \mu \zeta^1 \tag{5.16}$$

$$\zeta^1_t + \zeta^1_{xx} + u\zeta^1 = 0 \qquad (5.17)$$

by analogy with the K-P-D equation.

BIBLIOGRAPHY

1. F.B. Estabrook and H.D. Wahlquist, J. Math. Phys. 16, 1, (1975).

2. F.B. Estabrook and B. Harrison, J. Math. Phys. 12, 653 (1971).

3. B. Kadomtsev and V. Petviashvili, Soviet Physics Doke 15, 539 (1970).

4. V. Dryuma, J.E.T.P. Lett. 19, 387 (1974).

5. D. Benny and A.C. Newell, J. Math. Phys. 46, 133 (1967).

6. D. Benny, Nonlinear Wave Motion, A.C. Newell, ed., A.M.S., Providence, R.I., 1974.

7. R. Miura, Studies in Appl. Math. 53, 45 (1974).

8. P. Kulish and E. Nissimov, Pis'ma v JETP (Russian) 24, 247 (1976).

9. H.C. Moriis, J. Math. Phys. 17, 1870 (1976).

10. V. Zakharov and A. Shabat, Func. Anal. Appl. 8, 226 (1974).

11. H.C. Morris, J. Math. Phys. 17, 1867 (1976).

12. H.C. Morris, Prolongation Structures and Nonlinear Evolution Equations in Two Spatial Dimensions: A Generalized Nonlinear Schrodinger Equation, J. Math. Phys. (in press).

13. M. Ablowitz and R. Haberman, Phys. Rev. Lett. 35, 1185 (1975).

14. V. Zakharov and A. Shabat, Soviet Physics JETP 34, 62 (1972).

A PLAUSIBILITY ARGUMENT FOR THE MARCHENKO EQUATION

Alwyn C. Scott
The University of Wisconsin
Madison, Wisconsin

Consider the Sine-Gordon equation in characteristic (light cone) coordinates

$$\phi_{xt} = \sin \phi \quad (1)$$

This equation is of interest in several diverse areas of applied science [1]. Recently, Ablowitz, Kaup, Newell and Segur [2] have shown that the evolution of (1) from suitably prescribed initial conditions can be exactly obtained through a succession of linear calculations known as the "inverse scattering transform method" (ISTM) [2,3]. The gist of this method is as follows. Two linear operators (L and B) have been found which depend parametrically on $\phi(x,t)$ and for which the operator equation

$$iL_t = BL - LB \quad (2)$$

implies (1). Then the eigenvalues λ in

$$L\psi = \lambda\psi \quad (3)$$

remains constant with time if ψ evolves as

$$i\psi_t = B\psi \quad . \tag{4}$$

Thus, the ISTM can be carried through in three steps.

1) <u>Direct problem</u>. Given the initial data $\phi(x,0)$, consider it as a "scattering potential" for the operator L. The scattering parameters (i.e., the eigenvalues and eigenfunctions of (3)) are calculated at $t = 0$.

2) <u>Time evolution of the scattering data</u>. Since the eigenvalues remain constant with time, time evolution of the eigenfunctions can be calculated using (4) at large values of x (where ϕ is equal to some asymptotic value).

3) <u>Inverse problem</u>. From a knowledge of the scattering data at large values of x and as a function of t, $\phi(x,t)$ is reconstructed using the techniques of inverse scattering theory.

For the Sine-Gordon equation (1), the scattering equation (3) is

$$\psi_{1,x} + i\lambda\psi_1 = -\frac{1}{2}\phi_x\psi_2 \tag{5a}$$

$$\psi_{2,x} - i\lambda\psi_2 = +\frac{1}{2}\phi_x\psi_1 \tag{5b}$$

Assuming the reflection coefficient,

$$\rho(\lambda,t) = \lim_{x \to +\infty} \left(\frac{\psi_2}{\psi_1} e^{-2i\lambda x} \right) \qquad (6)$$

is known, (5) can be inverted as follows. First the kernel

$$F(\cdot) \equiv \frac{1}{2\pi} \int_{-\infty}^{\infty} \rho(\lambda,t) \exp[i\lambda(\cdot)] \, d\lambda \\ - i \sum_{j=1}^{N} c_j(t) \exp[i\lambda_j(\cdot)] \qquad (7)$$

is computed where the λ_j are UHP poles of ρ and the c_j are corresponding residues. Next the (Marchenko-type) integral equation

$$K(x,y) = F(x+y) - \int_x^\infty \int_x^\infty K(x,\alpha) F(\alpha+\alpha') F(\alpha'+y) \, d\alpha \, d\alpha' \qquad (8)$$

with $x-y < 0$ is solved for $K(x,y)$. Finally,

$$\psi_x = 4K(x,x) \quad . \qquad (9)$$

The point of this note is to present a plausibility argument for Equations (7)-(9) following a suggestion by Whitham ([4], p. 612) and the discussion in [3]. This approach to the inverse scattering problem has been discussed in detail by Balanis [5]. Although nothing substantial is added to the careful treatment by [3], this development may help the reader to feel more comfortable with the inverse scattering calculation.

The scattering Equations (5) can be written in the form

$$\Psi_{1,x} - \Psi_{1,y} = -\frac{1}{2}\phi_x \Psi_2 \qquad (10a)$$

$$\Psi_{2,x} + \Psi_{2,y} = +\frac{1}{2}\phi_x \Psi_1 \qquad (10b)$$

where y is a "pseudo time" defined through the Laplace transform

$$\Psi_i(x,y) = \frac{1}{2\pi} \int_C \psi_i(x,\lambda) \exp(-i\lambda y) \, d\lambda \quad , \quad i=1,2 \qquad (11)$$

In order to ensure causality in pseudotime, the curve c must lie above all the UHP poles of the reflection coefficient ρ.

Next <u>assume</u> that Ψ_1 and Ψ_2 can be represented in the form

$$\Psi_1(x,y) = \Psi_1^\infty(x,y) + \int_x^\infty [K_{11}(x,\alpha)\Psi_1^\infty(\alpha,y) + K_{12}(x,\alpha)\Psi_2^\infty(\alpha,y)] \, d\alpha$$
$$(12a)$$

$$\Psi_2(x,y) = \Psi_2^\infty(x,y) + \int_x^\infty [K_{21}(x,\alpha)\Psi_1^\infty(\alpha,y) + K_{22}(x,\alpha)\Psi_2^\infty(\alpha,y)] \, d\alpha$$
$$(12b)$$

where

$$K_{ij}(x,y) \to 0 \quad \text{as} \quad y \to \infty \qquad (13)$$

Substitution of (12) into (10) together with the assumption

$$K_{12}(x,x) = -K_{21}(x,x) = \frac{1}{4} \phi_x \qquad (14)$$

implies that the K's must satisfy

$$\frac{\partial K_{11}(x,y)}{\partial x} + \frac{\partial K_{11}(x,y)}{\partial y} = -\frac{1}{2} \phi_x K_{21}(x,y) \qquad (15a)$$

$$\frac{\partial K_{12}(x,y)}{\partial x} - \frac{\partial K_{12}(x,y)}{\partial y} = -\frac{1}{2} \phi_x K_{22}(x,y) \qquad (15b)$$

$$\frac{\partial K_{21}(x,y)}{\partial x} - \frac{\partial K_{21}(x,y)}{\partial y} = +\frac{1}{2} \phi_x K_{11}(x,y) \qquad (15c)$$

$$\frac{\partial K_{22}(x,y)}{\partial x} + \frac{\partial K_{22}(x,y)}{\partial y} = +\frac{1}{2} \phi_x K_{12}(x,y) \qquad (15d)$$

With the initial conditions (14), these can be satisfied for

$$K_{11}(x,y) = K_{22}(x,y) \qquad (16a)$$

$$K_{12}(x,y) = -K_{21}(x,y) \quad . \qquad (16b)$$

Next we choose

$$\Psi_1^\infty = \delta(x+y) \qquad (17a)$$

$$\Psi_2^\infty = F(x-y) \qquad (17b)$$

and note that the causality condition for (10) requires both $\Psi_1(x,y)$ and $\Psi_2(x,y)$ to be zero for $x+y < 0$. Then Equations (12) become

$$K_{11}(x_1 - y) + \int_x^\infty K_{12}(x,\alpha) F(\alpha-y) \, d\alpha = 0 \tag{18a}$$

$$F(x-y) + K_{12}(x - y) + \int_x^\infty K_{22}(x,\alpha) F(\alpha-y) \, d\alpha = 0 \tag{18b}$$

Using (16), (18) imply

$$K_{12}(x,y) = F(x+y) - \int_x^\infty \int_x^\infty K_{12}(x,\alpha) F(\alpha+\alpha') F(\alpha'+y) \, d\alpha' \, d\alpha \tag{19}$$

Since $F(x-y)$, as defined in (17), is an impulse response, it is the Laplace transform (defined in (11)) of the reflection coefficient, ρ. Thus,

$$F(x+y) = \frac{1}{2\pi} \int_C \rho(\lambda,t) \exp[i\lambda(x+y)] \, d\lambda \tag{20}$$

Assuming simple poles of ρ, Cauchy's integral formula (for the closed path formed by c and the real axis of the λ-plane) gives (7). Then (19) and (14) are equivalent to (8) and (9).

Final comment. The key step in the above development was the assumption stated in (12). From (15) and (16), it was possible to find a set of interaction kernels, the K's, from which the scattered waves could be represented in terms of their asymptotic forms. Development of ISTM's for equations with more than one space variable should require a corresponding property for the evolution of scattered amplitudes in pseudotime.

BIBLIOGRAPHY

1. A. Baroul, R. Esposito, C.J. Magee, and A.C. Scott, Riv. Nuovo Cimento 1, 227 (1971).

2. M.J. Ablowitz, D.J. Kaup, A.C. Newell, and H. Segur, Phys. Rev. Lett. 30, 1262 (1973).

3. M.J. Ablowitz, D.J. Kaup, A.C. Newell, and H. Segur, Stud. Appl. Math. 53, 249 (1974).

4. G.B. Whitham, Linear and Nonlinear Waves, Wiley-Interscience, New York, 1974.

5. G.N. Balanis, J. Math. Phys. 13, 1001 (1972).